CONTENTS

STATEMENTS

	Page
Batt, Hon. Philip E., Governor, State of Idaho	17
Carlson, Hon. Arne H., Governor, State of Minnesota	20
Craig, Hon. Larry E., U.S. Senator from Idaho	7
Gates, Gary, Vice President for Nuclear Operations, Omaha Public Power District, on behalf of the American Public Power Association	56
Grams, Hon. Rod, U.S. Senator from Minnesota	12
Johnston, Hon. J. Bennett, U.S. Senator from Louisiana	4
Kyl, Hon. Jon, U.S. Senator from Arizona	6
McNeill, Corbin A., Jr., President and CEO, PECO Energy Company	73
Murkowski, Hon. Frank H., U.S. Senator from Alaska	1
O'Leary, Hon. Hazel R., Secretary, Department of Energy, accompanied by Dr. Daniel Dreyfus	25
Pirie, Robert B., Jr., Assistant Secretary for Installations and Environment, Department of the Navy	50
Sherman, William K., on behalf of the Vermont Department of Public Service, and the Northeast High-Level Radioactive Waste Transportation Task Force	63
Snell, Dick, Chairman, President and CEO, Pinnacle West Capital Corporation	68
Wellstone, Hon. Paul, U.S. Senator from Minnesota	13

APPENDIX

Additional material submitted for the record 83

NUCLEAR WASTE POLICY ACT OF 1995

THURSDAY, DECEMBER 14, 1995

U.S. SENATE,
COMMITTEE ON ENERGY AND NATURAL RESOURCES,
Washington, DC.

The committee met, pursuant to notice, at 9:35 a.m. in room SD–366, Dirksen Senate Office Building, Hon. Frank H. Murkowski, chairman, presiding.

OPENING STATEMENT OF HON. FRANK H. MURKOWSKI, U.S. SENATOR FROM ALASKA

The CHAIRMAN. Let me wish you a good morning. We will open today's Energy and Natural Resources Committee hearing. The topic today is the Nuclear Waste Policy Act of 1995, S. 1271.

I am very pleased to welcome our distinguished group of witnesses today as we address the ultimate disposition of our spent nuclear fuel and the promises made by our Government to address that issue with some dispatch.

This is the second of our hearings on the disposition of spent nuclear fuel. In the last hearing, we heard a lot of criticism of the Department of Energy's nuclear waste program. We have a significant prop today. It is a piece of rock from Yucca Mountain. It actually marks progress: 1 mile of tunnel reached ahead of schedule, December 12, 1995.

I am told that they have drilled about 2 miles of the tunnel. I see some smiling faces out there that are nodding up and down, so that is a good sign. If any member of the committee questions whether progress has occurred, we have some evidence of that progress.

Now, I guess it is good news and bad news that they have dug 2 miles of the tunnel; there is 3 more miles to go. Yucca Mountain has been practically the sole focus of the Department of Energy program for about 12 years, and about $4.2 billion of the taxpayers' money. Now, it is true that all of that money was not spent at Yucca Mountain, but at least another $4 billion, we are told, or thereabouts, will have to be spent before two threshold events can occur. These are significant events.

One event is the Department of Energy's site suitability decision and the second is the NRC licensing application and approval. It will be interesting to see if we can get an update on what the odds of licensing and suitability might be, and whether there has been a lessening of the chance of either of those happening, which of course would question the advisability of having spent the $4.2 billion or thereabouts that we have spent on the project.

Now, a little bit about the Department of Energy's existing schedule. As I understand it, DOE's site suitability decision is in 1998. The NRC license application is filed in 2001, and waste acceptance in 2010. Now that is assuming that appropriation levels do not cause the schedule to slip.

Now, we have had strong objections to this program from Nevadans. As a matter of fact, some have suggested they are willing to lay down in the tracks of progress. I guess that is their right. Today, we will hear from Senator Bryan and Senator Reid, who have asked to testify. We have heard from Nevada before on this subject. This is not the last time we are going to be addressing this subject.

I do want the record to note a letter from Senator Bryan and Senator Reid and Bob Miller, the Governor of Nevada, indicating their concern over the bill. They state that the bill, in their opinion, is flawed, and as the proposed ultimate destination for the nuclear power industry's waste, Nevada will face, by far, the greatest impact. That is not a matter of dispute.

However, they go on to say that despite the importance of this legislation to Nevada, no Nevada witness has been invited or allowed to testify at the hearing today. I quote:

> For reasons unknown to us, tomorrow's hearing focuses on testimony related to relations between the Navy and the State of Idaho, a minor, if not totally irrelevant, concern in a bill focusing on the commercial nuclear waste issue.
>
> Accordingly, we request that the committee schedule additional hearings to provide an opportunity for extensive input from witnesses representing the State of Nevada prior to any committee action on high-level nuclear waste legislation. For too long, the debate over high-level nuclear waste has focused on the interests of the nuclear power industry, and ignored the interests of those most affected—namely, the citizens of Nevada.

I do not know that I agree that this action today is irrelevant, but I can assure my friends from Nevada that they will be heard, they have been heard, and we are very sensitive, of course, to the role that Nevada would play as a potential site for the repository.

Furthermore, I think it is fair to say that one of the major purposes of this hearing is to give Idahoans a chance to make their case, which really has not been done before. Today we will focus on the problems caused by the buildup of spent nuclear fuel around the country, and the necessity of having a solution. Again, I would like to remind you that we are aware of the intensity of emotion in Nevada. We saw the lawsuit filed by the Nevada Attorney General against Lincoln County officials for responding to our request for their opinion. I understand that the State tried to remove them from office for offering what we considered to be constructive suggestions.

Now, current law authorizes only a permanent repository at Yucca Mountain. And that is dependent on appropriations. The tunnelling will continue. The future of the permanent repository will eventually be resolved, but, of course, not in time to allow the Government to meet its obligation to accept delivery of spent nuclear fuel.

Also, Yucca Mountain is intended to handle disposal of spent fuel from research reactors, defense plants, Navy propulsion plants, at INEL, Hanford, Savannah River, Oak Ridge, universities, and

S. Hrg. 104-402

NUCLEAR WASTE POLICY ACT OF 1995

HEARING

BEFORE THE

COMMITTEE ON ENERGY AND NATURAL RESOURCES UNITED STATES SENATE

ONE HUNDRED FOURTH CONGRESS

FIRST SESSION

ON

S. 1271

TO AMEND THE NUCLEAR WASTE POLICY ACT OF 1982

DECEMBER 14, 1995

Printed for the use of the
Committee on Energy and Natural Resources

U.S. GOVERNMENT PRINTING OFFICE

23-370 CC WASHINGTON : 1996

For sale by the U.S. Government Printing Office
Superintendent of Documents, Congressional Sales Office, Washington, DC 20402
ISBN 0-16-052462-8

COMMITTEE ON ENERGY AND NATURAL RESOURCES

FRANK H. MURKOWSKI, Alaska, *Chairman*

MARK O. HATFIELD, Oregon
PETE V. DOMENICI, New Mexico
DON NICKLES, Oklahoma
LARRY E. CRAIG, Idaho
BEN NIGHTHORSE CAMPBELL, Colorado
CRAIG THOMAS, Wyoming
JON KYL, Arizona
ROD GRAMS, Minnesota
JAMES M. JEFFORDS, Vermont
CONRAD BURNS, Montana

J. BENNETT JOHNSTON, Louisiana
DALE BUMPERS, Arkansas
WENDELL H. FORD, Kentucky
BILL BRADLEY, New Jersey
JEFF BINGAMAN, New Mexico
DANIEL K. AKAKA, Hawaii
PAUL WELLSTONE, Minnesota
HOWELL HEFLIN, Alabama
BYRON L. DORGAN, North Dakota

GREGG D. RENKES, *Staff Director*
GARY G. ELLSWORTH, *Chief Counsel*
BENJAMIN S. COOPER, *Staff Director for the Minority*
KAREN HUNSICKER, *Counsel*
SAM FOWLER, *Chief Counsel, Minority*

other sites, and nuclear waste from weapons repository facilities at INEL, Hanford, Savannah River, and West Valley.

The Department of Energy also accepts foreign research reactor fuel for storage at Savannah River, and unfortunately we are committed, in the framework agreement to find a place to store fuel from North Korea once that eventuality comes about. Finally, Idaho's citizens have placed their trust in an agreement with the Navy that the Naval spent fuel stored at Idaho will be removed by the year 2035.

I think this is a leap of faith. We have the ratepayers' concerns, who have trusted the Government's word that spent fuel would be taken away in the beginning of 1998. They feel they have performed their part of the deal by paying nearly $11 billion into a nuclear waste fund. However, it is painfully clear that the Government is not prepared to perform its part of the deal when 1998 comes about.

In response, this bill, S. 1271, would authorize an interim storage facility at the Nevada test site. This is certainly not an easy, noncontroversial approach. Senator Craig's bill represents one approach. There are other ideas as well. Other nations, such as France, England, and Japan, reprocess spent fuel. The result, of course, is a dramatically reduced volume of waste. Some scientists have proposed promising technologies for the elimination of nuclear waste after reprocessing.

Interim storage would let us take our time investigating other options, including a full study of the Yucca Mountain site. It would also allow us to keep our minds open and pursue other options as changes in science, changes in economics, and changes in politics as well, make them feasible.

In the past, these issues have been addressed in somewhat of a piecemeal fashion. Although there has been an awful lot of hard work and dedicated effort that has gone into the process, we have failed to prioritize and allocate resources in the most cost-effective way. I think we are all aware of that. Resources have, to some extent, been mismanaged and the parties involved, including multiple agencies, boards, academies, and Congress, all seem to blame each other.

Problems with multiple authorities and a lack of accountability have hurt our efforts to clean up DOE's weapon sites as well. With both of these programs, I think we need to concentrate authority. That is not going to be easy, but some say the only way to get it done is under one super-agency that will have the responsibility for these duties.

This is not, certainly, a partisan matter. Senator Johnston has worked long and hard to reach an accord on this. But the reality is that if we fail, we will just have one more problem that all of us in Government—Congress, the administration—Republicans and Democrats—have failed to solve.

Now, the administration has yet to provide a proposed solution. That is why we look forward to hearing the ideas of Secretary O'Leary today.

I would hope to investigate all of these options as the committee works towards the development of a chairman's mark that will start us on a path to dealing with these issues in a comprehensive and long-term way. And I believe that we should do this in the 104th Congress, notwithstanding the pervasive influence of presidential politics that will occur during this year.

I certainly look forward to the testimony of the witnesses today.

PREPARED STATEMENT OF HON. JAMES M. JEFFORDS, U.S. SENATOR FROM VERMONT

Mr. Chairman, I want to thank you for holding these important hearings. I would also like to welcome Bill Sherman, the State of Vermont's Nuclear Engineer and Chair of the Vermont State Nuclear Advisory Panel. Bill is responsible for oversight over the Vermont Yankee Nuclear Power Plant, based in the southern part of Vermont. He's a resident of our capital city, Montpelier, Vermont, and is here today to stress how important it is that during this Congress we move to resolve some of the difficult issues related to nuclear waste.

I think Bill Sherman and I would both agree that this hearing is the beginning of a process which should result in the development of a safe and cost efficient interim storage facility for our nation's spent nuclear fuel. I would like to commend Senator Craig for introducing S. 1271, legislation which should move us in this direction.

Many opponents of this type of solution are simply hoping to stop any future development of new nuclear power plants. This view is shortsighted. What we should primarily be concerned with is the health and safety of the residents of the regions where this waste is currently being stored. Regardless of your position on nuclear power, temporary storage of spent nuclear fuel at over 100 plants across the country is not acceptable. I believe that Congress must act in an expedited fashion to solve this problem and I am committed to passage of legislation which will provide for safe disposal of nuclear waste.

I want to thank Bill Sherman for joining us today to educate us on how we can safely transport this nuclear waste and I want to assure Senator Craig that I will do all I can to help him in his efforts to move his legislation.

Thank you.

PREPARED STATEMENT OF HON. RICHARD H. BRYAN, U.S. SENATOR FROM NEVADA

Dear Mr. Chairman, given the importance of the legislation to be considered by the Committee on Energy and Natural Resources during tomorrow mornings hearing to the State of Nevada, we are very disappointed with the failure of the Committee to invite witnesses representing the interests of the citizens of Nevada.

S. 1271 is a seriously flawed bill, and places the health and safety of all Americans at risk. By mandating unprecedented, and unnecessary, shipments of high-level nuclear waste throughout the country, nearly every state will be negatively impacted by its passage. As the proposed ultimate destination for the nuclear power industry's toxic waste, however, Nevada will face, by far, the greatest impacts.

Despite the importance of this legislation to Nevada, no Nevada witnesses have been invited, or allowed, to testify at tomorrow's hearing. For reasons unknown to us, tomorrow's hearing focuses on testimony related to relations between the Navy and the State of Idaho—a minor, if not totally irrelevant, concern in a bill focusing on the commercial nuclear waste issue.

Accordingly, we request that the Committee schedule additional hearings to provide an opportunity for extensive input from witnesses representing the State of Nevada prior to any Committee action on high-level nuclear waste legislation. For too long, the debate over high-level nuclear waste has focused on the interests of the nuclear power industry, and ignored the interests of those most effected, the citizens of Nevada.

The CHAIRMAN. Senator Bennett Johnston, please.

STATEMENT OF HON. J. BENNETT JOHNSTON, U.S. SENATOR FROM LOUISIANA

Senator JOHNSTON. Thank you, Mr. Chairman.

Mr. Chairman, that which masquerades as America's nuclear waste policy is nothing less than a disaster, an outrage of astonishing dimensions. It is a waste of money of a proportion that every taxpayer of this country ought to feel outraged about.

We have an administration unwilling to submit a bill, not only unwilling to submit a bill, unwilling to submit a plan, unwilling to communicate honestly with the Congress and the American people about what is involved in this issue. They frustrate their own people. Dr. Dan Dreyfus, who is a very capable man, is trying to work to fashion a plan, but the administration completely frustrates its own people, submits no plan to the Congress, does not tell the American people how they are going to deal with this, and treats it as a political problem between now and the next presidential election.

In the meantime, the Congress is not doing anything either. I submitted a bill here nearly a year ago, after having worked with Dr. Dreyfus in the Nuclear Waste Policy Office, and have been trying to get that bill up and cannot do it. Over in the House, they have not passed legislation. The clock is ticking. It is costing an enormous amount of money each year.

It is going to cost utilities some $5 billion between 1998 and 2010 to store their own waste—$5 billion. That is in addition to the 1 mill per kilowatt hour under which we have collected $10 billion and spent $5 billion. Now, this is not chicken feed. It is going to cost them $5 billion. They have already collected $10 billion. They are going to collect a lot more money. And it is all for naught. No plan. No bill. No action.

Now, I really believe we need interim storage, and we need to get to it as quickly as we can, because it is going to cost probably more than $500 million a year for every year we delay. We have already delayed a whole year in this committee. We have delayed a whole year. The House has delayed a whole year while this issue was very important and needed tending to. And what is that year going to cost? Probably over $500 million. That is not chicken feed.

And what do we do? Well, I put in a bill. I put in a provision in the appropriations bill, the energy and water appropriations bill, to deal with this, to get started with long-lead-time items on interim storage. And what did the Congress say? Let the authorizing committees deal with it.

Well, I hope the authorizing committees are ready to deal with it. There has been a year delay. So, look, there is plenty of complicity in this, what I call a disaster of a non-policy, a disaster masquerading as a policy.

Now, Senator Craig has submitted, I think, a good bill. I find some problems with it which need to be worked out. And I believe Senator Craig and I can work on this with other members of the committee. Let me just identify some of the problems I find with the bill, which I think need to be worked out. It has got a 1998 date that is not even remotely achievable. And the problem with that is it is all right to have 1998 as a goal, but if you say you can get sued if you do not meet the goal and have to take the waste, that gives you a budget act point of order.

Now, we have got to get real in this committee and in this Congress and propose and pass legislation that can get enacted into law. We are not looking for an issue. We are not out to try to make utilities sort of happy with today's Pyrrhic victory only to find that there is a point of order that prevents the legislation from being passed tomorrow. But we can work on that.

We got a billion-dollar railroad route provided in the Craig bill. We need to study that. There are cheaper routes. Do we really want to spend that billion dollars? And can we have a billion-dollar railroad, continue payments to the States, and interim storage, and all within putting a cap on the nuclear waste feed? These are serious, serious questions. But these are things we can talk about, along with putting this program off budget in some way.

Nobody has submitted a plan, I do not believe, except in the legislation I have proposed, to get this program funded. You have got a $10 billion fund; $5 billion has been spent; $5 billion is collected from the ratepayers—that is, the taxpayers of this country—to provide for this issue. And yet you cannot spend it because it comes out of the allocation of the energy and water appropriations subcommittee, where everything else in the country, all kind of competing priorities, need to be funded. And as a practical matter, you just cannot get at that.

Now, we need to spend the money collected for the purpose of dealing with nuclear waste—for the purpose for which it was collected. And we need to deal with that openly and honestly and do it. Otherwise, there is no solution to the problem.

I believe that people in this country need to demand of this administration and to demand of this Congress that we solve the problem. It is not a technically impossible problem. It is a challenge. But it is well within our scientific capability to solve this problem safely and with some dispatch. It is only a question of political will on the part of the administration and on the part of the Congress.

Now, we need to solve the problem which we know how to solve. It is not difficult to solve on any basis other than a political basis.

Mr. Chairman, this is the year—at least starting in January—is the year that we need to solve this problem finally, once and for all, and deal safely with America's nuclear waste problem.

The CHAIRMAN. Thanks, Bennett.

Senator Kyl.

STATEMENT OF HON. JON KYL, U.S. SENATOR FROM ARIZONA

Senator KYL. Mr. Chairman, just let me ask consent to put my statement into the record, and to agree with the statement that Senator Johnston has just made. I think it well summarizes the sorry state of affairs in which we find ourselves today. And I am very much looking forward to being able to do something about this problem. To know the amount of money that has just been spent on site characterization, it is over $2.5 billion, I think, just on site characterization and what do we have to show for it, is to just suggest one small aspect of this problem.

So I think we ought to get on with the hearing, but I certainly agree that we have got to do our part to get this problem resolved.

[The prepared statement of Senator Kyl follows:]

PREPARED STATEMENT OF HON. JON KYL, U.S. SENATOR FROM ARIZONA

Thank you, Mr. Chairman, for holding this hearing today. I look forward to working with you and other members of the Committee on moving this legislation to resolve this important national problem. Sound management of spent reactor fuel and high-level defense waste is vital for the health and safety of the citizens of this country. I believe that the urgent problem of what to do with our nuclear waste is one that can be solved, if only the Department of Energy would step forward and take responsibility. DOE has a commitment to accept spent fuel for storage starting in 1998. The Nuclear Waste Policy Act of 1995 is needed for our nation's energy security; to safeguard the environment; and, most importantly, to fulfill a commitment the federal government made to the nation's electricity consumers long ago.

Despite clear policy direction from the Congress, the programs of the Department of Energy to develop interim and permanent storage facilities for a spent fuel and high-level defense wastes have proven both costly and ineffective. There are a variety of reasons behind this lack of progress, ranging from technical questions in the site characterization process to political and legal delays at the state level. The most recent estimate from the Department indicates the permanent repository will not be ready to accept spent fuel until, at the very earliest, the year 2015.

In my state of Arizona, electric customers have paid approximately $150 million into the Nuclear Waste Fund since Palo Verde, the nation's largest nuclear energy plant, began operating less than 10 years ago. All told, customers nation-wide pay more than $600 million a year into this fund. Since 1982, when the fund was established by an Act of Congress, Americans have sent more than $11 billion to the federal treasury for the purpose of dealing with the nation's spent nuclear fuel.

Unfortunately, the problem today is worse than ever. The money collected for spent fuel disposal has not been spent for its intended purpose, which was to provide storage for spent nuclear fuel beginning in 1998: Over $4.7 billion has been spent out of the Fund, $2.7 billion of that on site characterization studies for a deep geologic repository at Yucca Mountain, but as I mentioned earlier, the DOE has indicated this storage facility cannot become operational until 2015, at the earliest. Despite the Department's clear mandate to transfer spent fuel from the nation's 109 nuclear plants beginning in 1998, there is currently no central location in which to store the spent fuel.

The legislation which we are considering today puts the program back on track. Specifically, the bill:

- Establishes a central storage facility large enough to accommodate 100,000 metric tons of the nation's spend fuel—enough holding capacity until the deep geologic repository is completed;
- Will not raise fees paid by customers into the Nuclear Waste Fund, and still makes interim storage available in 1998;
- Provides a benefits package to the state hosting the central interim storage facility;
- Allows for use of existing NRC-licensed technology for the transportation and storage of used fuel; and
- Establishes a transportation system to move spend fuel from reactor sites to the central interim storage facility.

It is important to note that this legislation would allow continued progress on the permanent geologic repository at Yucca Mountain—it does not throw up any roadblocks. DOE can go down both paths at the same time. As a matter of fact, Secretary O'Leary and I discussed yesterday that DOE was following a similar dual-track regarding tritium production. I don't see any inconsistencies here.

It is imperative that Congress take action now to force the Department of Energy to meet its contract with the American people to develop a facility and begin accepting spent fuel from utilities in 1998—just three years from now. It's an obligation we accepted at the beginning of the age of nuclear energy and that we reconfirmed back in 1982. To fulfill this obligation, we must pass this vital legislation.

Thank you again, Mr. Chairman, for holding this important hearing. I look forward to hearing from the witnesses.

The CHAIRMAN. Thank you, Senator Kyl. Of course, that is why we brought this rock, to ensure that we could show some progress.

Senator Craig.

STATEMENT OF HON. LARRY E. CRAIG, U.S. SENATOR FROM IDAHO

Senator CRAIG. Mr. Chairman, at the outset, let me thank you for agreeing to hold this hearing on my legislation, S. 1271. And, Mr. Chairman, let me also say that this is not an Idaho issue. This room would not be full of citizens today if it were just an Idaho issue. This is a major national problem that this committee must deal with, and we must deal with it now. And I can only say, as it relates to Senator Bennett Johnston's comments, right on, Senator, right on. You are absolutely correct that this has to be now faced by us, and faced quickly and appropriately.

And I think Senator Johnston is correct: The different between his legislation and mine are not obstacles at all. They are problems that can be worked out and will be worked out.

I would also say, Mr. Chairman, last night, I met with several of the House members. They have the votes to pass nearly identical legislation in the House at this moment. And it is very likely, before we go out of session, before Christmas, that they will pass that legislation. So the clock is not just ticking, it is running, and it is running pretty loudly.

One of the reasons that this room is full is because of the gentleman who sits in front of us this morning at the witness table, my Governor from Idaho, Phil Batt. Idaho may be a mouse, but this mouse roared over the last couple of years, and said some very loud things to this country. And it is not just an agreement, Mr. Chairman, that might be met in the year 2035, it is a Federal court order that will stop all things, including drydocking the U.S. Navy, if this Congress does not face up to its responsibilities and deal with it in an appropriate fashion.

That is part of why I have introduced the legislation I have. I was 4 years old when nuclear waste began to arrive in my State of Idaho. And Governor Batt, doing what he had to do as a responsible guardian of the public trust in Idaho, dealt straightaway with a lawsuit, in trying to bring the Department of Energy and the U.S. Navy into what we believed was a workable solution to a problem that Idaho and this Government and this Congress has ignored for some time.

We have 10,800 shipments of spent nuclear fuel and waste in my State, sitting over a major aquifer that needs to be dealt with in a timely fashion, and our Governor has spelled that out with the use of a Federal court, and the Navy and the DOE have agreed.

So, in response to that, Mr. Chairman, it is our Government today that has agreed, as they agreed with legislation passed in 1982, with the commercial segment of the nuclear industry in our country, to begin to receive nuclear waste. And Bennett Johnston is absolutely correct: That is a joke. It is a law, but it is a legal joke. We are nowhere near able to do so, and we must. And that is an obligation that we will accept and we need to accept in a timely fashion.

Mr. Chairman, I will not, for the sake of time, go into all of my statement. I would ask unanimous consent that it become a part of the committee record this morning.

The CHAIRMAN. Without objection.

Senator CRAIG. But, H.R. 1020, the House legislation that I have spoken of, and my legislation, or something very similar to that, needs to make it to this President's desk in a timely fashion. And the Secretary of Energy needs to be a part of a final solution, and certainly I and I know that Senator Johnston is willing to work with the Department and the Secretary, Dan Dreyfus, and others who have to be and should be major players in a solution.

But politics as usual will not stand on this issue anymore. We are going to have it on the floor of the Senate, and we are going to debate, and we are going to debate it loud and clear. And we are going to push this to a final solution. I think Senator Johnston and I are committed to that, and it just simply must get done.

My legislation does deal with the development of an interim facility to be ready by 1998. Is that practical? Yes, it is, if we started tomorrow or we started in 1996. We have done amazing things in this country when we wanted to do it. To pass it off as something that is incapable of being done, when we have already spent $2 billion and are prepared to spend $4 billion more on drilling holes in rocks in large mountains in Nevada, tells me that we have tremendous capability if we choose to do so.

I am also suggesting that we do it in two phases that are approachable. From handling 20,000 metric tons in the first phase, to 100,000 in the second phase, the development of a transportation system that works for the short-term and the long-term. And while Bennett Johnston and I and the State of Nevada might argue over the route—and that is certainly an arguable point—the reality is that the transportation system has to be developed, and we ought to get on with the business of developing it, and doing it as quickly as possible.

My legislation in no way slows down the effort at trying to create the kind of site preparation and suitability work that is going on at Yucca Mountain. They ought to go hand in glove. We are talking about interim, and we are talking about permanent. And both of them ought to be, and must be, done. And I think that is what our American public wants at this moment to deal with this issue.

The Department of Energy must purchase transportable storage containers and take advantage of the technology that is available. One of the games—and I repeat this—one of the games that has been played is the idea that you cannot use existing technology, you always have to reinvent it when we deal with this issue. And that takes years and years longer.

We also know, too, Mr. Chairman, that we have been some unrealistic tests before ourselves, and we did it for political purposes—the 10,000-year test—that kind of stuff that is almost humanly unachievable. But we know we can build interim surface dry storage now that is just as safe and sound for the American concern as any that we are capable of, and that engineering talent is available today.

So there is no need for delay. What there is, is need for political will, for political action. I am extremely proud to have our Governor, Phil Batt, with us today, because he took on what needed to be done in the State of Idaho. He had the political will and the courage to do something that was not necessarily very popular, Mr. Chairman. Because what Nevada is doing today is what is popular, and that is the no, hell no game.

Our Governor chose not to do that. He had to deal and face an issue that was real and before us to protect our environment, our people, and at the same time to recognize a national responsibility. He has assumed that. We expect other States to assume the same. But, more importantly, we expect this Congress and this administration to accept and assume that same responsibility.

So, Mr. Chairman, I am extremely pleased to have our Governor here to testify today.

Thank you.

[The prepared statement of Senator Craig follows:]

PREPARED STATEMENT OF HON. LARRY E. CRAIG, U.S. SENATOR FROM IDAHO

Mr. Chairman, I want to thank you on behalf of myself and the 18 other Senate cosponsors of the Nuclear Waste Policy Act of 1995 for scheduling this hearing today. It is only through the passage of S. 1271 that we will finally provide for the timely storage and disposal of spent nuclear fuel and high-level nuclear waste from the nation's defense program and commercial nuclear power plants.

I would also like to welcome Governor Phil Batt from my home state of Idaho who will testify on this bill today. It has been a pleasure to work with Governor Batt as he crafted a historic agreement between the State of Idaho, the U.S. Navy and the Department of Energy. Shipments of spent nuclear fuel began accumulating at the INEL when I was a child growing up in Midvale, Idaho, in 1949 and continue to this day. However, until Governor Batt signed an agreement in 1995, there was no provision to remove this material from Idaho. I am proud to have worked with him to help to craft the agreement that assures liquid wastes will be put into dry form to protect the Snake River aquifer and approximately 10,800 shipments of spent nuclear fuel and transuranic wastes will begin to be shipped from Idaho in 1999 and be completely removed by 2035.

To help to accomplish this objective, the Nuclear Waste Policy Act of 1995 creates an integrated system that will ensure construction of an interim storage facility and permanent repository to manage the legacy of America's great defense force, including spent fuel from the Navy's nuclear-powered fleet of aircraft carriers and submarines, currently stored in my state of Idaho, as well as components from dismantled nuclear weapons and commercial spent fuel from about 73 sites in more than 34 states.

Mr. Chairman, transferring nuclear waste from the many defense and commercial nuclear sites to a single federal facility beginning in 1998 was the intent of Congress when it passed the Nuclear Waste Policy Act of 1982.

Unbelievably, we are only three years from the date when the Energy Department is obligated to begin accepting this radioactive waste, and the DOE is still studying a site in Nevada to determine if it is a suitable location for a deep geologic repository for high-level radioactive waste. Because of endless bureaucratic delays that have plagued the program, the federal government now says it will not have a repository operating until 2010, at the earliest, and probably several years later than that.

That is 12 years after the federal government is contractually obligated to take title to spent fuel from civilian power plants and more than 10 years after the people of Idaho were first promised that high-level waste stored at the Idaho National Engineering Laboratory would be moved to a permanent repository.

Mr. Chairman, INEL has now managed spent nuclear fuel from Navy warships for more than 30 years. More recently, it has also become the resting place for spent fuel and other radioactive components from Three Mile Island. Like many nuclear facilities across our nation, INEL has served the federal government and the citizens of America well. But now, the federal government must accept its responsibility under law to take nuclear waste to a facility licensed by an independent regulator where it can be managed safely and economically.

The bill I am introducing authorizes construction of a federally-licensed facility on the Nevada Test Site near Yucca Mountain to store spent Navy fuel from Idaho National Engineering Laboratory and other defense facilities and spent fuel currently stored at commercial nuclear power plants from Maine to California. The bill instructs the federal government to begin operation of an interim storage facility in 1998 so that high-level radioactive materials can be transferred to the Test Site, where it can be more easily managed.

Transferring nuclear materials from sites around the country to a single facility holds several advantages over the current system. First, because the interim storage facility provided in my bill will be licensed by the Nuclear Regulatory Commission ensuring safe storage of all materials. Second, a single site will be far more economical to maintain and keep secure. Finally, the storage site designated in my bill is close to Yucca Mountain, the likely site of a permanent repository for high-level waste.

Mr. Chairman, though some will surely disagree, I do not think it is unreasonable to assume that Yucca will eventually be judged as suitable for a permanent repository. Nor do I think that establishing a storage site near the mountain compromises the integrity of the ongoing scientific studies.

It is important to recall that scientists and engineers at Yucca Mountain have conducted the most thorough and comprehensive geological survey ever undertaken on any piece of earth. After $5 billion in expenditures, the scientists have found no reason why the site would not be suitable for a permanent nuclear waste repository. Moreover, S. 1271 ensures that research at Yucca Mountain will continue during construction and operation of an interim storage facility.

My bill is similar to legislation (H.R. 1020) that passed the House Commerce Committee 30–4 on August 2 and awaits House floor action. My bill includes the following provisions that reform the federal government's spent fuel management program in these critical areas:

- The bill reaffirms the federal government's responsibility to begin accepting waste from defense and commercial nuclear facilities in 1998;
- Authorizes construction of an interim storage facility in two phases with date-certain schedules. Phase one will allow acceptance of up to 20,000 metric tons of uranium (MTU), including defense program waste, and phase two permits up to 100,000 MTU.
- Authorizes the Energy Department to develop a transportation system to move spent fuel safely from America's defense and commercial nuclear facilities to this single storage site.
- Authorizes continued development of a permanent repository program according to DOE's 1994 "program approach."
- Requires the Energy Department to take title to spent nuclear fuel at plant sites and to operate a transportation system from a contract holder's designated site(s) to a federal interim storage facility.
- The Energy Department must purchase transportable storage containers, taking advantage of technologies available in the market-place. Defense spent fuel must be transferred to containers that can be used at a storage facility licensed by the Nuclear Regulatory Commission.
- Funding priorities for the Energy Department's program should be (1) interim storage and a related transportation system, (2) Construction of a railroad spur in Nevada from existing rail lines to the interim storage facility and (3) scientific study for a repository location.

The principle difference between the House bill and my bill revolves around future funding for civilian spent fuel management. The House Committee voted to change current law, which has resulted in the federal government collecting more than $11 billion from utilities and their ratepayers over the last 13 years, while spending less than half of that amount for the purpose it was intended—that is, building a nuclear waste repository. The rest of the money, more than $5 billion, has been used to finance our deficit spending habit.

The House bill ensures that in future years appropriations in any given year will equal contributions from ratepayers. If Congress votes to reduce funding for the program, collections from utilities and ratepayers will be similarly reduced.

My bill retains the current funding mechanism for the DOE program. I hope as we proceed in the Senate, however, that we will take a close look at the House funding provision or something similar to help ensure that Congress once and for all moves toward ending the practice of collecting funds for specific purposes and then using them to help balance our swollen budget.

Mr. Chairman, this legislation will solve an important issue for the citizens of Idaho . . . and for all Americans. The question of how best to manage spent nuclear fuel and other radioactive materials has been considered for most of my lifetime. There is no question that centralized storage and disposal in a remote location is better than leaving nuclear waste scattered across the United States. It is time to implement a centralized storage program and to develop the solution that protects public health and safety and the environment.

A dozen years ago, the federal government signed contracts with utilities and agreed to take title to and dispose of used nuclear fuel by 1998. Now it is time for the federal government to live up to its commitment to these consumers and to the residents of states like mine who have played an essential role in managing the waste from the nation's nuclear defense program.

There is widespread support for these principles among state governors, attorneys general, utility regulators and more than 180 members from both sides of the aisle in the House of Representatives and 18 members in the Senate. I urge my colleagues to support this legislation to manage the nation's nuclear waste in an integrated, sensible fashion and to demonstrate to the American people that the federal government can honor its commitments.

The United States has benefitted from the many uses of nuclear materials, whether as a deterrent to global conflict or nuclear fuel that is used to generate electricity in a manner that emits no air pollution or other emissions harmful to our environment.

Our generation has benefitted enormously from these diverse uses. We have enjoyed peace, economic prosperity and a clearer environment. Now, our generation must finally take responsibility to properly manage spent fuel from the defense program and from more than 110 commercial nuclear power plants.

I will work to assure this bill moves forward in a timely fashion.

The CHAIRMAN. Thank you, Senator Craig.

Senator Campbell.

Senator CAMPBELL. I have no opening statement, Mr. Chairman.

The CHAIRMAN. Senator Grams.

STATEMENT OF HON. ROD GRAMS, U.S. SENATOR FROM MINNESOTA

Senator GRAMS. Thank you, Mr. Chairman, and I want to commend you for holding today's hearing on our Nation's nuclear waste policy. It is my understanding, as Senator Craig has mentioned, that the House plans to act on similar legislation before the Christmas recess, and I believe today's hearing will provide the impetus for timely action in the Senate. And I also wish to associate myself with the remarks made by Senator Johnston and Senate Craig about the timeliness of getting this bill passed or something worked out in the very near future.

When this committee held a hearing on various nuclear waste policy proposals earlier this year, all of our witnesses, including Minnesota's Public Utility Commissioner, Chris Sanda, agreed that the 1998 date is critical in this debate.

And now, with 1996 only a few weeks away, the deadline is rapidly approaching. And right now today, we are no closer to resolving this issue than the last time Congress enacted nuclear waste legislation back in 1987.

But it is not like we have not seen this deadline coming. For 16 years, the Department of Energy has been charged with the responsibility of our civilian spent fuel. And in that time, DOE has spent nearly $5 billion of ratepayers' money, including over $250 million from Minnesota's electric consumers. And yet, here we sit debating the issue of exactly what to do with America's civilian nuclear waste.

But the Department of Energy just continues to go around in circles. First, they said that we can store waste at Yucca Mountain. Then they tell us we cannot force it on Nevada. Then DOE says they cannot meet the 1998 deadline, and even claim they are not legally bound to do so. And then they tell us they can, if only Congress would untie their hands. The latest was that an interim facility could not be completed for 7 years, at a cost of nearly $400 million. And then their testimony says it could be done in 4 years at a third of the cost. Well, it is hard to tell what is truth and what is fiction for the DOE anymore, and it is my hope that today's hearing will help us find out more.

For the past few months I have met with and carefully reviewed reports and studies by the General Accounting Office, independent groups, current and former DOE employees, and even former Energy Secretary Watkins on this issue. And I have found that there is strong support for removing the civilian waste program from

DOE, and that support grows even stronger when we focus specifically on privatizing the interim storage program.

Now, after years of working on this issue with Minnesota's ratepayers, utilities, and State officials, I am convinced that privatizing the interim storage program remains our last and best hope for getting waste out of Minnesota and the other 30-plus States which are struggling with this issue as well.

Late today, I will be introducing the Private Interim Storage Facility Authorizing Act of 1995, and my legislation targets one small but key component of the overall Federal nuclear waste program.

Privatizing the interim waste storage program offers three key benefits. First, it saves money. It provides relief to States, such as Minnesota, which are faced with on-site storage restrictions, and it also protects the environment.

And as with most initiatives, privatizing the interim storage program would improve efficiency and lower costs to the taxpayers. And based on the business plan for the Mescalero Private Initiative, a private interim facility could be completed for approximately $135 million, and done in time to meet the 1998 deadline.

Now that is hundreds of millions of dollars less than what the Office of Civilian Radioactive Waste Management originally quoted earlier this summer. And while DOE has revised its estimates downward in recent weeks, it is only because of the undisputable cost data for the Muscalero project.

Now competition always reduces costs, particularly when it is with the Federal Government. The Muscalero project is proof of that. With the Muscalero initiative moving forward while congressional action has stalled, some have asked if this venture could resolve the interim use issue. The short answer would be yes. And in fact, 20 utilities, 20 utilities are now participating in this private venture.

But Congress and the DOE have the legal responsibility to resolve this issue, and we have the responsibility of ensuring that it be done by 1998. I believe focusing on the interim storage privatization not only represents a workable option, but it also provides a win-win for ratepayers, and for States and for the environment.

So, Mr. Chairman, in closing, I look forward to hearing the testimony of our distinguished witnesses today, and look forward to working with you, with Senator Craig, with Senator Bennett, specifically on the privatization issue in the coming months. And I yield back the balance of my time.

The CHAIRMAN. Thank you, Senator Grams.

Senator Wellstone, good morning.

STATEMENT OF HON. PAUL WELLSTONE, U.S. SENATOR FROM MINNESOTA

Senator WELLSTONE. Thank you, Mr. Chairman.

The CHAIRMAN. We are not talking about ANWR this morning, so you can feel right at home.

Senator WELLSTONE. I am sure that we will have ample opportunity to talk about the Arctic.

The CHAIRMAN. It is a done deal. We put it together last night. [Laughter.]

23-370 - 96 - 2

Senator WELLSTONE. I do not think the Arctic Refuge will ever be a done deal.

The CHAIRMAN. Well, it will be when you come up.

[Laughter.]

Senator WELLSTONE. Mr. Chairman, in Minnesota, I am sure as my colleague has just said, nuclear waste is a huge issue. And we know how important it is to deal with the problem. The trick is to deal with it responsibly. I do not believe, I say to my good friend from Idaho, that S. 1271 meets that criteria. And I would have thought that we would be starting to get the message that moving away from basic protection of the environment is opposed by the American people. The bottom line is that Americans do not want to see environmental laws trashed.

Mr. Chairman, that being said, I wonder why we are considering a bill that explicitly limits the application of the National Environmental Policy Act and, after naming a few Federal statutes, provides that actions under the bill, "shall not be subject to any other Federal, State, or local Atomic Energy, environmental, or land-use laws, regulations, or orders."

What ever happened to the historic Congress of block grants and States rights? And why is judicial review of agency action under the law curtailed? Why does the bill remove Federal employment protections for certain DOE personnel?

Mr. Chairman, we all know why. S. 1271 is a classic case of overreaching, an example that I think we can put on the shelf next to all the other ones we have seen this year. The bill introduced by my colleague from Louisiana, a bill that I could not support, cannot support, apparently just did not go far enough for the nuclear industry. It was not enough to ship nuclear waste to Nevada before we knew if Yucca Mountain is capable of hosting a permanent repository, the bill now before us would send the waste whether or not a facility exists there to accept it.

Mr. Chairman, it is remarkable that a Congress that claims to pride itself on protecting the taxpayer could so ignore the tremendous taxpayer liability embodied in this bill. I brought this up at the last nuclear waste hearings in this committee, and I repeat it now: If you allow DOE to accept title to the waste—and that is exactly what this legislation calls for—you are passing on a potentially infinite liability to generations of taxpayers. That is exactly what you are doing by transferring title. This is no small issue, believe you me. And when people in the country find out about it, they will have a lot to say.

The wonderful irony of this is that passing the liability to the taxpayer has been the law since 1982, and there has been an understanding even longer. S. 1271, which we are here today to discuss, actually recites the apparently empty finding that, "the costs of such storage and disposal should be the responsibility—this is a direct quote—of the generators and the owners of such waste and spent fuel."

Mr. Chairman, let me just finish by laying some numbers out here. In 1990, DOE was estimating the total life cycle costs for a single repository and interim facility combined would be $25.6 billion, and that the fund would ultimately have taken in $19 billion by the time nuclear utilities closed their plants and stopped paying the fee. That is a $6.6 billion shortfall.

Now, DOE's estimate of the total life cycle cost is $33.1 billion. But there has been no change in the amount of money utilities are expected to pay into the fund. And where the 1990 cost estimate of $25.6 billion included both an interim facility and a permanent repository, the new $33.1 billion cost estimate does not even include an interim facility.

So the costs are going up. The revenue is remaining static. This is a surefire formula for shortfall.

Now, Mr. Chairman, let me just go to the kicker. There is a shortfall, but now, last but not least, the kicker. Despite all of the evidence that the numbers do not add up, S. 1271 takes the unprecedented step of capping the fee at the current level of 1 mill. So I guess there is not going to be a need for the DOE to do an adequacy report, since the Secretary will no longer have the authority to do anything to address the shortfall.

Why, will somebody please tell me, is it necessary to tie the Secretary's hand in this way?

And do not forget who is going to be on the hook, everybody, when the fund provides insufficient funds in the long run. The taxpayers in this country—because we are transferring the liability to them—will be on the hook because the taxpayer will own title to the waste over the next 10,000 years. That is the beauty of this deal for the nuclear industry. They fight now to force DOE to take the waste that they themselves have generated, knowing full well that if they succeed they will be off the hook.

Mr. Chairman, this looks like the savings and loans bailout, with a little bit of a toxic twist.

Mr. Chairman, I announced my position early on—went through my principles. DOE should be directed to construct and begin loading waste into one or more interim facilities by January 31, 1998. No question about it. But when they do this, these facilities must meet applicable standards to protect the public health and safety, be as economic as possible, and be paid for out of the nuclear waste fund.

Mr. Chairman, I could go on about other such work that needs to be done. I think I have made my point clear. You cannot waive the environmental standards. You cannot tell States you are going to supersede their rules and some of their standards. You cannot tell people who are worried about how this is going to be transported and whether or not their families might be put in some jeopardy that there is not going to be anything they can do. And you cannot transfer this liability to the taxpayers—as is done in this legislation. This piece of legislation is severely flawed, and I think there will be a major battle in the country over this one.

Other than that, I do not feel strongly about it.

The CHAIRMAN. Well, that is fine.

[Laughter.]

Senator CRAIG. Mr. Chairman, I think he is opposed to my bill.

[Laughter.]

The CHAIRMAN. Well, we have had an opportunity to hear some broadly varying views, and we have a bill before us that attempts to make some constructive corrections. The Senator from Minnesota has highlighted some of his concerns, as well as concerns expressed by others. We seem to have spent, over the time that this has been before us, an awful lot of energy on this issue. We have not been able to deal with it under existing environmental laws, for reasons that we all have some knowledge of.

We are faced with the extraordinary dilemma that nobody wants nuclear waste. So if interim storage is going to happen, it is going to have to be forced on someone. Obviously that is not a very palatable solution.

This committee would look rather foolish if indeed we completed the Yucca Mountain tunnelling and basically got that repository up to a point where it could either accommodate retrievable or nonretrievable storage—permanent or temporary, depending on which avenue you want to pursue—and found that we had expended some $8 billion to $10 billion of the taxpayers' money and found it unsuitable. Based on a decision in the 1998 time frame, I think the last testimony before this committee suggested that we had about an 80 percent chance of the site meeting suitability standards.

Further, in the year 2001, when the license application would be completed, the odds given to this committee at one time for licensing were 50/50.

Now, to expend the amount of dollars that would be required to put Yucca in that time frame, and yet find that the odds are not any better than that, one wonders about the responsibility that we have as members of the committee to continue under the current environmental oversight program that the Senator from Minnesota has discussed. Unfortunately, it just seems like we simply cannot get there from here. And the question is: What solution can we find to address this problem?

The Blush report, which Senator Johnston worked on, to give you some idea, found that we had expended $7.5 billion on the cleanup of Hanford so far. We really have not started the cleanup process there, and the "guesstimate" is somewhere in the area of up to $200 billion will have to be spent, if you are going to clean it up and put it back in its natural state. And the question is: Is that feasible? There are just going to have to be some tough decisions made, recognizing the responsibility of government, as well as the private sector, to meet this demand.

The issue of access into the Yucca site is something I have spent some time on as well. I am not personally convinced it is necessary that a billion dollars be spent on a railroad. There are other methods and other technologies that might be much cheaper and still do the job.

So, in spite of the energy of the committee, this is not something that is going to be done overnight, because there are a lot of concerns, a lot of considerations, and a lot of responsibilities. And I want to ensure that when we have something, it is going to work and it is going to become a reality, and it is not just going to be another dry run up a blind canyon.

So, with that, Senator Craig, would you care to introduce your Governor, and we will proceed with the hearing.

Senator CRAIG. Well, thank you very much, Mr. Chairman. I am extremely pleased today to have the committee hear from our Governor, Phil Batt, of the State of Idaho. As I mentioned in my opening comments, Governor Batt, when first taking office, was faced with dealing with a lawsuit brought by our former Governor, Cecil Andrus, to cause an environmental impact statement to be done over our National Engineering Laboratory, the INEL, in southeastern Idaho, as to a variety of aspects of management at that facility, but primarily waste management and how it would go forward and what could be anticipated.

And you and I and this committee knows the work of the Department and what is proposed and needed and recommended for many of our laboratories to receive. It was upon that action, under a Federal court action, that our Governor began a negotiation.

We are the sole repository of spent naval fuel rods. And that has been a heritage that Idaho has borne proudly, because it not only lent to its national defense participation, but it also produced a nuclear navy much more efficient today than it was when it was first put to sea. But it was also a problem that our Governor had to face and did face for the sake of Idaho, our environment.

The Senator from Minnesota needs to know that I and this Governor are terribly concerned about our environment, as that Senator is, and that he has badly portrayed the legislation before us. But we will let the Governor speak to Idaho's needs and the importance of finding interim storage to meet not only a Federal court order in Idaho, but, as I have said, the responsibility of this country to its citizenry.

Governor Batt, welcome to the committee. We are pleased to have you here.

Mr. Chairman.

The CHAIRMAN. Governor, you have waited long and we have waited to hear you, so it is your turn.

STATEMENT OF HON. PHILIP E. BATT, GOVERNOR, STATE OF IDAHO

Governor BATT. Well, thank you, Chairman Murkowski, members of the committee, Senator Craig. It is a high honor for me to be able to visit with you this morning and give you my views of why S. 1271 should become law.

I would like to parenthetically say that I really appreciate Senator Johnston's remarks—they pretty well mirror mine—in that we have the wherewithal to come to grips with where we are going with nuclear waste now. We have the money to do it. We have the science to do it. We lack the political willpower. And I do not mean that in a partisan way. We just have to come to grips with this problem.

As I have said many times in my home State, this Nation entered the nuclear era with a woeful lack of planning for the disposal of our nuclear byproducts. We must come to grips with this problem, and cut through the political rhetoric and open these depositories, both the waste isolation pilot plan in New Mexico and a permanent or interim repository near Yucca Mountain, Nevada.

Past Congresses and past administrations have made attempts to address this issue, but they have been unwilling to face the music and achieve a final solution to our nuclear waste dilemma. I am proud to see that this Congress appears willing to tackle this issue with an eye toward definitive action. I am pleased to see that a member of your committee, Idaho's senior U.S. Senator, Larry Craig, is tackling the spent nuclear fuel problem head-on, with the introduction of S. 1271. I applaud Senator Craig and this committee for the bold steps you are taking on behalf of this Nation.

It is unfortunate that it has taken us so long to arrive at this point. Senator Craig's legislation, S. 1271, is important to Idaho because of the large amounts of spent nuclear fuel and other nuclear waste already in our State.

Idaho started down the road of nuclear research in 1949, when 850,000 acres of Idaho land was named the National Reactor Testing Station—it was later renamed INEL. As a result of that early effort, on December 21, 1951, Idaho became the first place in the world where electricity was produced from nuclear power. Since that time, 52 nuclear reactors have been built in Idaho. Nearly all of those reactors have been research reactors that have furnished mankind's knowledge of the atom. And I would like to say that none of these produce commercial electricity.

Research done in Idaho has helped the U.S. Navy develop nuclear reactors that now last more than 20 years. And since the inception of the Naval Reactor Program in the 1950's and until this year, hundreds of U.S. Navy personnel were shipped to the high plains of Idaho to learn how to run nuclear reactors.

With the Navy's high-tech, cutting-edge scientific research and training done in my State, our Navy has been able to maintain a strong military presence throughout the world. Our nuclear Navy is truly one of the premier success stories in the history of warfare. I am proud that Idaho is part of that distinguished heritage.

Idaho's nuclear research has had other military applications as well. Armor for the M–1 Abrams tank and military vehicles is made in Idaho with depleted uranium. I have been told that the armor built in Idaho saved American lives during the Gulf War, and I am proud of that fact.

Research done in Idaho has also helped the commercial reactor community as well. When the Three Mile Island reactor went down, the reactor core was sent to Idaho for examination. And Idaho's scientists were able to verify the problems that the reactor went through. It was Idaho's scientists who were able to make recommendations to prevent such an accident from happening again.

But while Idaho is famous for our potatoes, we are also famous for our contributions in advancing nuclear science, and I am proud of all that. But for all our great successes, nuclear waste has accumulated in Idaho by the ton, all the time with promises that it would some day be removed.

The day I took office, over 261 metric tons of spent nuclear fuel and nearly 2 million gallons of liquid radioactive waste were already in Idaho. Tons of other nuclear waste have accumulated over the decades as well. Over the past 40 years, over 4,000 truck equivalent shipments of radioactive material have come to Idaho for, "temporary storage." The first concerns about these shipments started at least 2½ decades ago.

In 1970, a former U.S. Senator from Idaho, Frank Church—a member of the opposite political party I might say—received a letter from the head of the U.S. Atomic Energy Commission, which is now the DOE. He was told the nuclear waste would be removed from Idaho within the decade. In 1973, former Idaho Governor Cecil Andrus received assurances from Federal officials that the "nuclear waste in Idaho would begin to be removed within 10 years."

25 years later, no nuclear waste has left Idaho.

In 1982, Congress passed the Nuclear Waste Policy Act. Transuranic waste was supposed to go to WIPP by 1988. Spent fuel was supposed to go to Yucca Mountain in 1998.

Many deadlines have come and gone, and the waste still remains. WIPP will not now open until 1998—or will not open in 1998 unless you pass legislation now before Congress to open it a year earlier. And it looks like Yucca Mountain now will not open for several decades.

Because of these broken promises, my predecessor initiated a lawsuit against the Federal Government in an attempt to keep further waste from coming into Idaho. I successfully concluded those efforts when I signed an historic agreement, which has been incorporated into a Federal court order that requires nearly all spent fuel and transuranic radioactive waste to be removed from Idaho by a specific date. Idaho now stands alone as the only State in the Nation with a Federal court order which requires nearly all nuclear waste to leave by a date certain.

I am proud of that achievement.

Until I signed the agreement, there was no legal requirement that nuclear waste ever would leave Idaho. Until I signed the agreement, my State may have been destined to become a de facto nuclear waste permanent repository. Until I signed the agreement, there was no commitment to move the Navy's fuel to an interim or permanent repository.

Idaho's situation is unique, as Senator Craig mentioned, because the spent fuel and other nuclear waste in Idaho are located over the Snake River Plain Aquifer, one of the largest aquifers in the world. We depend on that for our lifeblood in the State of Idaho. Removing this waste from above the drinking water of thousands of people is an absolute necessity. That is why it is so critical to Idaho that you open either Yucca Mountain or, as S. 1271 calls for, an interim repository near Yucca Mountain.

If an interim or permanent repository is not opened up in Nevada in the next few years, then the Department of Energy and Navy will have to begin looking at other DOE labs or other locations for a place to ship material which is now stored in Idaho. That material cannot stay perched above our aquifer past the deadlines agreed to by our Federal court decree.

The position I have taken today to open permanent geologic repositories is supported by the Western Governors' Association. Although the WGA has not yet taken a position on S. 1271, the various members of the WGA, both Republicans and Democrats, have over the years repeatedly called for the opening of permanent geologic repositories for this Nation's spent nuclear fuel and high-level waste.

Indeed, the Western Governors unanimously passed a resolution I sponsored earlier this year which calls for congressional and presidential, "action for the permanent disposal of the Nation's high-level radioactive waste."

But since we are still decades away from getting Yucca Mountain open, it is now necessary to look at other options. That is why an interim repository near Yucca Mountain makes so much sense. That is why S. 1271 is needed now. Its passage will cause the opening of an interim repository near the only place in the Nation which has been considered for a national repository for spent fuel.

Spent nuclear fuel must go someplace. And for Idaho, a Federal court order says it must be removed from its current location above our aquifer.

It is now up to this Congress to solve this Nation's looming nuclear waste crisis. It will take a bipartisan effort. It will require statesmen. It will require decisive leadership. For far too long we have kicked this problem down the road. It is now time to solve it.

So my message to you is simple. Get a repository opened up. Do it safely. Do it quickly. It is time to finish the job.

Thank you, Mr. Chairman.

The CHAIRMAN. Thank you very much, Governor. We certainly appreciate your statement and understand your sensitivity to the problem because, as you pointed out, promises were made and time has gone by and you still have the waste. And obviously, it is the responsibility of the Federal Government to meet that commitment. So we look forward to working with you and would appreciate your continued input and suggestions as we develop a proposal that will relieve you of the responsibility that was not created by the State of Idaho, yet nevertheless is a legacy that you have to live with.

First of all, let me for the record indicate that Governor Arne Carlson of Minnesota has asked me to express his regret that he was unable to be here, but we will place his testimony into the record as if read.

[The prepared statement of Governor Carlson follows:]

PREPARED STATEMENT OF HON. ARNE H. CARLSON, GOVERNOR, STATE OF MINNESOTA

Chairman Murkowski, members of the Committee, thank you for this opportunity to present testimony on the bills before you concerning the U.S. civilian nuclear waste program. This U.S. Department of Energy program is of vital interest to the State of Minnesota and to our nation. Changes are needed now to ensure its ultimate success, thus protecting us from unnecessary future expenses that could run into tens of billions of dollars—and to avoid the prolonged storage of already existing nuclear waste at 73 sites in 34 states.

CRITICAL JUNCTURE

Our nation's civilian radioactive waste program is at a critical juncture. It desperately needs program changes to provide a centralized, temporary radioactive waste storage facility by 1998, to facilitate waste transportation, and to continue a program to provide a permanent repository in the longer term.

Congress, including this Committee, is presently considering several bills to reform and fund the civilian radioactive waste program. The need to reform this program is now widely recognized. However, even a reformed program cannot be conducted effectively if electricity consumers' money in the Nuclear Waste Fund is diverted for other purposes. I am very concerned about a possible lack of Congressional resolve to carry this program to completion as promised.

PROMISES TO KEEP, CONTRACTS TO PERFORM

Nuclear energy was promoted in part by the Federal government's promise to remove radioactive waste from power plants. Congress decisively committed the Federal government to take and dispose of civilian radioactive waste through the Nuclear Waste Policy Act of 1982, and its amendments in 1987. The U.S. Department of Energy (DOE) further committed the Federal government through administrative proceedings establishing the rules which set contract terms under which radioactive waste would be taken, stored and disposed of. Other administrative proceedings, such as the Nuclear Regulatory Commission's Waste Confidence Rule, told the American public that we could literally bank on the Federal government's promise. Finally, the U.S. DOE signed individual versions of the Standard Contract with all civilian nuclear utilities required to dispose of civilian high-level radioactive waste, committing DOE to begin waste removal by 1998.

Despite the U.S. Department of Energy's public denial of its obligation this past summer, the Federal government has repeatedly committed itself to take title to high-level radioactive waste from civilian electric power plants and to transport it to storage and disposal facilities. Electric utilities, and state and local governments made multi-million dollar decisions relying on the full faith and backing of the Federal government that these promises would be kept and the contracts performed.

THE MINNESOTA EXPERIENCE

Minnesota hosts three utility nuclear reactors—one at Monticello and two at Prairie Island—owned and operated by Northern States Power Company. These plants provide 30 percent of Minnesota's electricity needs, are extremely reliable, provide emissions-free electricity, and are the state's lowest-cost electricity producers.

The interior holding pool for used nuclear fuel at the Prairie Island plants (PI) ran out of storage space in 1995. Without additional storage, Northern States Power (NSP) would have been forced to close this power plant, removing 20 percent of its generating base from service. To prevent this premature shutdown, NSP proposed to construct and operate a dry-cask storage facility, whereby the oldest and "coolest" waste would be removed from the pool, placed in large metal casks, stored outside the power plant in a secured area. Following Minnesota laws, NSP sought a certificate of need from the Minnesota Public Utilities Commission (MnPUC) for its proposal.

This project was subjected to intense review over and above the license review conducted by the Nuclear Regulatory Commission. State agencies in Minnesota performed an environmental impact statement, need and cost reviews, and assessments of alternative projects that could meet the state's energy needs. Of major concern throughout the proceedings was the viability of the federal program: Would the waste generated in Minnesota ever be removed and disposed of by the federal government, as required by federal law? Would the $210 million already paid by Minnesota ratepayers ever result in a facility that met their needs? Despite these questions, after reviewing all of the evidence, our Public Utilities Commission (MnPUC) granted limited approval of NSP's request. The rationale for limiting approval was to match Minnesota's storage capacity with DOE's obligation to begin removing waste: granting more capacity would remove needed pressure on DOE to meet its obligation for disposal, and granting less capacity would remove the state's lowest-cost supply of electricity from service before acceptable alternatives could be developed.

Groups opposing the project—including the Prairie Island Mdewakanton Sioux Community and a coalition of environmental organizations—appealed the MnPUC's decision. The appeal sought a ruling that the project required approval from the Minnesota Legislature before it could proceed. The Court agreed with these parties, finding that—because there was no assurance as to when the federal government would begin removing waste from Prairie Island—the waste may remain in Minnesota permanently, thus requiring legislative approval.

The Minnesota Legislature considered NSP's proposal in its 1994 session. The issue was the most difficult, divisive, and contentious issue facing the Legislature for years. Despite evidence demonstrating that the project was safe, that rates would increase by up to 17 percent without the project, and that environmental damage would increase were Prairie Island replaced with other sources of electricity, approval of the project was by no means certain. In the final hours of the session, the Legislature ultimately approved the project, subject to large increases in renewable energy statements and a search for a new Minnesota site to store the waste. The Legislature's decision, however, is only a temporary solution to the problem, and all parties expect that the issue will be brought back to the Legislature, perhaps even this year.

The situation in Minnesota—while contentious and dramatic—is in no way unique. Instead, this situation will be repeated throughout the country as other nuclear plants run out of storage space if the federal government has no place to put it. The Congressional Quarterly has estimated that by 1998, the number of power plants needing additional storage to continue operation will expand from the handful today to 26. Within the following three years the number will expand to 82. Given the 3 to 5 year lead times to create new nuclear waste storage facilities, our nation is clearly at the threshold of spending billions of dollars that would not be necessary if the federal government were prepared to fulfill its obligations as promised and committed to contract.

Resolving the problems in the nuclear waste disposal program is essential to this nation: public policy which allows waste to continue to remain stranded at the country's 72 nuclear power sites is unacceptable. The nation's ratepayers have lived up to their obligations under the Nuclear Waste Policy Act to fund the disposal program—contributing $11 billion to date and continuing to pay at a rate of $600 million each year. For safety, cost, and policy reasons, the federal government must meet its obligation for nuclear-waste disposal. It is widely believed that legislative reform is necessary to ensure this.

THE NUCLEAR WASTE STRATEGY COALITION

As I watched our experience unfold in Minnesota, I became convinced that dramatic actions would be required to ensure that the federal government fulfilled its obligations for safe, timely, and cost-effective nuclear-waste disposal. Recognizing that Minnesota would be more effective acting with others, I directed my Public Service Commissioner Kris Sanda to seek assistance from other states with unique interests in solving the nuclear-waste problem. Together with Michigan and Florida, Minnesota founded the Nuclear Waste Strategy Coalition, a group of state regulators, executives from nuclear utilities, and Attorneys General who work together to move the program forward to successful resolution. Our growing Coalition now has participants from 18 states, including Alabama, California, Connecticut, Florida, Illinois, Indiana, Iowa, Massachusetts, Michigan, Minnesota, New Hampshire, New York, North Carolina, Ohio, Rhode Island, South Carolina, Vermont, and Wisconsin. By bringing together diverse parties—even traditional adversaries—who are committed to solving this problem, our Coalition has become an innovator and a leader in this debate.

The Coalition is following a three-pronged action plan to bring about needed changes in the program. These actions include: holding the federal government accountable for its obligations for nuclear-waste disposal by initiating a lawsuit against the Department of Energy; supporting and encouraging development of private storage facilities to meet pressing, immediate needs for away-from-reactor storage; and developing and proposing solutions to the problems in the current program.

Members of the Nuclear Waste Strategy Coalition were instrumental in bringing together the Mescalero Apache Nation and Northern States Power Company to explore development of a private storage facility for nuclear waste. That project involves nearly two dozen nuclear utilities who have now formed a corporate entity and begun the first phase of the project.

Members of our Coalition are also playing a leading role in the lawsuits pending in federal court against DOE by the states and the nuclear utilities. Forty state agencies in 29 states, 25 investor owned utilities, and 8 municipal utilities have joined in these suits, indicating the depth of concern with the program.

Although we do point out the flaws in the current program, we are also committed to finding solutions to developing an effective program that successfully meets the nation's pressing need for nuclear waste disposal. At my direction, Commissioner Sanda has extensively explored alternatives and prepared discussion papers to help foster debate.

SIX NECESSARY ELEMENTS OF REFORM LEGISLATION

Let me now turn to the 6 points critical to reforming the U.S. civilian radioactive waste program. To overcome past problems of the program's lack of public confidence, cost escalation, schedule lapses, and its difficulty in obtaining Congressional appropriation of ratepayer money from the Nuclear Waste Fund, 1995 legislation reforming the Nuclear Waste Policy Act must:

1. Begin Waste Removal in 1998—The federal government is unconditionally contracted to begin removing radioactive waste in 1998 from the 73 temporary storage sites now at nuclear electric power plants in 34 states. It is not sufficient to simply take title or possession of the waste, the federal government must begin to remove waste from power plant sites beginning in 1998.

2. Release Ratepayer's Money for Intended Purpose—The American public is right to expect that the ratepayer-funded Nuclear Waste Fund will be used to address nuclear waste and that Congress will appropriate the necessary money from the fund to do so. In the next year alone, electric ratepayers will pay over $600 million into the Nuclear Waste Fund. The United States government promised to use these funds to begin removing high-level radioactive waste beginning in 1998, and to provide for its permanent disposal.

Over $11 billion, including interest, has been paid into the Nuclear Waste Fund and $6 billion remain held in trust by the federal government. While it's admirable that Congress wants to solve the federal deficit problem, it is simply unimaginable to many that the 104th Congress would take ratepayer's money in the Nuclear Waste Fund for other purposes. This money was collected to provide safe, timely, and cost-effective storage and permanent disposal of civilian high-level radioactive waste. The American people are right to expect it will be released for this purpose, not kept to provide accounting camouflage for the federal deficit. Use of the Nuclear Waste Fund for other purposes is an unjust and fraudulent tax on the American electricity consumer.

3. Provide a Central Temporary Storage Facility—A temporary, centralized radioactive waste facility must be authorized, sited in Nevada, and funded in order to provide the United States with timely, safe, and cost-effective interim storage of radioactive waste. Congress must establish an aggressive waste acceptance schedule for storing waste in the interim facility. This facility will augment and facilitate our nation's permanent radioactive waste disposal program, not replace it.

4. Continue a Permanent Disposal Program—Characterization of the Yucca Mountain, Nevada site must continue. State governments, utilities, and the public have acted in reliance on the federal government's promise that waste would be removed from power plant sites beginning in 1998 and permanent disposal provided. To ensure that deep geologic disposal remains an essential program element, within budget constraints, the program must be redesigned to improve management structure, reflect program priorities and provide incentives for efficiency.

5. Facilitate Transportation—Authorize the designation, construction and operation of facilities to transport civilian high-level radioactive waste to a central temporary storage site and to a permanent disposal facility. Provide necessary transportation corridors and rights-of-way to ensure access to the designated temporary storage facility and the permanent disposal facility.

6. Cap the Nuclear Waste Fund Fee—Cap the Nuclear Waste Fund payments at the present one-tenth of a cent per kilowatt-hour to ensure that the program costs resulting from past performance problems of the federal government are not shifted to electricity consumers.

These six elements are needed in the final legislation reforming the Nuclear Waste Policy Act to protect continuing consumer investment in the Nuclear Waste Fund which already exceeds $11 billion, and to ensure that the federal government fulfills its obligations for the interim storage and permanent disposal of civilian high-level radioactive waste. Civilian high-level radioactive waste now stored at 73 power plants in 34 states must be addressed. Legislation is needed and the time to enact it is now.

COMMENTS ON BILLS PRESENTLY PENDING BEFORE CONGRESS

The 1995 Congress presents the window of opportunity for comprehensive reform of the nuclear waste program. Through the actions such as the states' and utilities' lawsuits against the Department of Energy, the private spent-fuel storage project, and through high-profile state cases like Minnesota's Prairie Island case, the continuing failure of the program run by the Department of Energy is becoming widely recognized. There is a widespread consensus on the need for Congressional action.

I wish to commend Senator Craig, Senator Grams and Senator Johnston for each introducing legislation that addresses important problems in the nuclear waste program. Their bold steps have brought this important issue to the forefront and fostered discussion in the Senate. In the House, the debate is being brought forward as well. H.R. 1020 introduced by Representatives Upton and Towns is hopefully nearing a floor vote as we speak. I believe it is now time for the Senate as well to advance comprehensive legislation before this historic opportunity slips away.

I recognize that there are powerful forces fighting to preserve the status quo—to do nothing. Some of these special interests suggest that we are asking you to rush to judgment. If the 12 years in which we have wrestled with this dilemma is not enough time to see what needs fixing, no amount of time will be enough. Other special interests wrongly and loudly assert that reforming the nuclear waste program is a "budget buster" and that Congress has yet to find a way to pay for it. Let me make it perfectly clear that electricity consumers in Minnesota and throughout the nation have already paid, and continue to pay for civilian nuclear waste storage and disposal. This vital program reform only affects the budget to the extent that planners had hoped to divert electricity consumers payments from the Nuclear Waste Fund to fund other spending purposes! Furthermore, the "scare" numbers that these opponents are waving completely ignore the fact that a properly reformed program would run far more cost effectively than the present program.

CONCLUSIONS

Legislation is needed to correct the problems and flaws in the present federal program and to ensure that the nation secures safe, timely, and cost-effective storage and disposal facilities. The U.S. civilian nuclear waste program, while making minor progress at great cost, is not meeting the needs of the nation. Decisive action is needed to correct fundamental flaws in the program. Congress must not miss this opportunity to do so.

The nation's electricity consumers must not be burdened with tens of billions of dollars in otherwise unnecessary expenses. Having already paid $11 billion for a program that continually fails to meet their needs, America's electricity consumers demand and deserve no less.

The CHAIRMAN. He indicated that he wanted to note his strong support for the legislation that would provide a safe, timely, and cost-effective centralized storage site, to paraphrase his remarks.

Senator Craig.

Senator CRAIG. Thank you very much, Mr. Chairman. Let me only make brief comments so that we can keep this hearing on time. But I do appreciate, Governor Batt, your response and your testimony and your outlining of promises made, promises broken that this Federal Government has done not just to Idaho but to States like Idaho. who have become by inaction and promises broken, de facto storage States, if you will, for spent nuclear fuel and nuclear waste. And while this Congress has acted, it has not had the will to push through its actions and to cause them to happen, as it can, largely because of all of the tripwires that it wants to put up, even when it knows what it does, is scientifically sound and humanly very safe and environmentally sound. And that is part of the difficulty we are being frustrated with today.

Several years ago before this committee, when we began to look at the exorbitant costs of Yucca Mountain, I remember, Mr. Chairman and Governor, factitiously saying why do we not just go ahead and build it, and if we do not like it and it does not work, throw it away? And that is marvelous hindsight now because we would have been a billion dollars ahead had we done so.

But we are not doing that. The time is necessary, and the court order that you have signed or caused to happen, Governor, is important, and let me congratulate you for advancing this issue nationally, because there is no question that the actions that the State of Idaho has taken under your leadership have pushed this to the national forefront again to cause it to come to resolution. And let me congratulate you for that.

Thank you very much for being here.

Governor BATT. Am I excused?

Senator CRAIG. You are.

The CHAIRMAN. Thank you, Senator Craig.

Any further statements?

[No response.]

The CHAIRMAN. Governor, I just have one question.

I assume that you are satisfied that Idaho does not hold title, in the strict legal sense, to the waste in your State.

Governor BATT. Well, usually I consult with my lawyers about that, but yes, I am convinced of that.

The CHAIRMAN. You are satisfied of that. Okay. So the title and responsibility really rest with the Federal Government?

Governor BATT. That is correct, sir.

The CHAIRMAN. Senator Johnston, do you have any questions?

Senator JOHNSTON. No. Thank you very much, Governor.

Governor BATT. Thank you.

The CHAIRMAN. Next we will hear from the Secretary of Energy, Secretary Hazel O'Leary, and I think Robert Pirie.

Secretary O'LEARY. As I understand it, Mr. Chairman, you requested Dr. Dreyfus, and I am very comfortable to have him with me.

The CHAIRMAN. That is fine. Dr. Dreyfus, we welcome you to the committee, and, Madam Secretary, you look very becoming in green, even though it is Christmas and not St. Patrick's Day. But around here we are not so sure, anymore.

Anyway, Madam Secretary. You have heard some of the statements, no doubt. You have heard the level of frustration that has been expressed, even anguish, by some from the Northern States. We look to you to provide us with the clear path to heaven. So we look forward to your presentation.

STATEMENT OF HON. HAZEL R. O'LEARY, SECRETARY, DEPARTMENT OF ENERGY, ACCOMPANIED BY DR. DANIEL DREYFUS

Secretary O'LEARY. Thank you, Mr. Chairman and members of the committee. As the chairman has indicated, I did have the benefit in the library of watching and hearing most of the opening statements this morning. And I understand clearly the frustration of this committee, certainly of Governor Batt and his insistence that the Department keep its commitment to remove all spent fuel from the site at Idaho National Engineering Laboratory by 2035, and I understand clearly the frustration of the industry, as well.

I recall the last time I appeared before this committee. We were grappling over some of the same issues. And while I understand and clearly recognize the goal and the purposes of S. 1271, I regret to inform this committee that our administration cannot support this legislation. And I would like to walk through very briefly why that is so.

My formal testimony has been filed, and I would spare you the reading of same, but like very quickly to come to the substance of our administration's disagreement.

The CHAIRMAN. Your statement will be taken into the record as if read.

Secretary O'LEARY. Thank you.

Not with the principle nor the outcome stated, but quite frankly with the very specific nature of the designation of an interim waste facility before the critical scientific and technical work is done to permit us to understand whether or not the repository permanently can be sited at Yucca Mountain in Nevada.

Our sense several. As the President has indicated in writing, we feel specifically that it is premature to point out Nevada as a site for an interim storage before we have clear certainty whether or not that facility will become the de facto above-ground storage facility. That is the very crux of the matter, and I have heard and I have known for some years the utter frustration of the Department of Energy not being able to meet its 1998 deadline.

The effort now is, of course, to move us forward precipitously to that deadline without the required and the requisite finding that will let us know what is, in fact, the future for the site now proposed to become the interim storage facility, will it indeed be the permanent de fact storage facility, which is much the dilemma that the Governor of Idaho expresses this morning.

We have a number of other concerns and considerations, as does the Congress, and I have heard the expression of many of the Senators here this morning with respect to funding and whether the funding will carry us through against the current design. I have also heard discussion about the fact that the Department has pushed back its program, if you will, Senator Grams, reduced some of the work that we believe was required in the early suitability assessment that we were to complete in order to meet some of the requirements that the Senate and the House has outlined in the appropriation bill for 1996. And I need to express that very briefly, and I will try to get off of it.

In essence, what is in the new budget mark for this program, which I think most of us would admit has been very successful in meeting its milestones in this past year. As a matter of fact, Dr. Dreyfus is about 43 or 45 percent ahead of schedule in terms of boring within the tunnel. That would not have been a thing I could have testified to the first year that I was here. Nonetheless, we have tried to reduce our program in order to continue the characterization of the mountain and get some of the work done.

The budget has been reduced for characterization in effect by 50 percent, slowing down and trimming up the technical data that we can provide to move forward with licensing, which application we had earlier known and committed to have before the NRC in 2001. Nonetheless, our view is that with the information we will have in hand in 1998, that is the more appropriate date to begin the consideration of the siting of an interim facility. At that point this committee and the Congress generally and the American people will know what are the conditions at Yucca Mountain, and will also know, as the Congress takes its action, which, of course, is appropriate to take. It will know what the future of that site is.

For this reason and a number of others, which I understand will be probed on question and answer, we regretfully inform you that our administration cannot and would not support this bill.

Thank you, Mr. Chairman.

[The prepared statement of Secretary O'Leary follows:]

Prepared Statement of Hon. Hazel R. O'Leary, Secretary, Department of Energy

INTRODUCTION

Mr. Chairman and members of the committee, I am pleased to appear before you today. I understand that the committee intends to utilize S. 1271, introduced by Senator Craig, as the basis for its markup of a high-level waste bill. Our administration is opposed to this legislation for reasons I will elaborate on in this statement.

The primary emphasis in this bill, as in the similar measure that has been reported by the House Commerce Committee, is to address the concerns of the institutions that are faced with extended at-reactor storage of spent fuel from commercial power plants. Their desire to provide for removal of spent fuel from reactor sites is well understood. By 2010, about 55 sites in 30 states will require new spent nuclear fuel storage capacity for over 11,000 metric tons of uranium (MTU).

There is a strong case against leaving spent fuel at 73 sites in 34 states for several decades or longer. As power plants complete their licensed terms and go out of service, we will have nuclear facilities operated for the sole purpose of spent fuel storage. This scenario would be "de facto" public policy to defer a decision on long-term management. Certainly, no rational case can be made that this is the preferred long-term management approach.

We are concerned, however, that the approach adopted in the proposed bill for resolving the interim storage need is likely to raise new issues for equally important aspects of the national nuclear waste system. The bill as drafted assigns priority to an immediate initiative on interim storage with a highly accelerated schedule. This requirement coupled with the probable budget constraints would place the repository program in jeopardy and reduce the policy commitment to the long-term strategy of geologic disposal.

Geologic disposal is the end-point of the federal radioactive waste management system. A decision to significantly defer or abandon it would have the effect of making at-reactor or interim storage of commercial spent fuel the open-ended, long-term strategy. If technical or institutional considerations were to prevent supplemental storage, economic and efficient generating capacity would be prematurely retired to the detriment of the ratepayers, the utility companies and the national energy outlook.

Geologic disposal also underpins our efforts to clean-up the radioactive residue of decades of nuclear weapons production. The Department has more than 100 million gallons of high-level radioactive waste that must be processed into tens of thousands of glass canisters requiring deep geologic disposal. Geologic disposal is the technical foundation for our international stance on nuclear nonproliferation and our advocacy of limiting the international trade in weapons materials. Our endorsement of the international consensus on geologic disposal sets an example for the high standards of environmental protection that we wish to promote worldwide.

Geologic disposal is also the strategy for the ultimate disposition of 65 MTU of spent fuel from nuclear powered naval vessels, about 2,700 MTU of spent nuclear fuel owned by our government, and 19 MTU of spent fuel of U.S. origin in foreign research reactors. If we turn away from this considered strategy and instead are left with a default strategy of perpetual managed storage for defense wastes, our objective of encouraging the highest possible global standards of nuclear safeguards and safety would be undercut.

Unless the importance of geologic disposal is reflected in the articulation of public policy and in the funding of the program, we will need to rethink our options for the long-term custody of all high-level nuclear waste.

CONCERNS WITH S. 1271

The objectives of S. 1271 are consistent with the long established Federal obligation to provide for the long-term custody of the nation's spent nuclear fuel and high-level radioactive waste. Many of the provisions address issues that have arisen through prior efforts to meet this obligation. However, we are concerned that an emphasis on establishing an interim storage facility in the vicinity of the Yucca Mountain site, and at an unreasonably rapid pace, will force this program into excessively curtailed regulatory processes and unreasonable and logistically untenable deadlines. These provisions could also invite unnecessary public opposition to the objectives of the program.

We are also concerned that, given the funding priorities for interim storage in S. 1271 and the likelihood that funding will be constrained there will not be sufficient funding to maintain an efficient schedule for the geologic disposal program. The Yucca Mountain project is the only process to determine if deep geologic disposal can attain public acceptance through an independent regulatory process in the United States. The results of that process are essential to the credibility of the commitment we have made to weapons system cleanup and probably will have important consequences for the options we are considering for the stabilization and elimination of weapons grade materials as well.

The bill's provisions mandate the immediate construction of a facility for the interim storage of commercial spent fuel. The facility is scheduled to begin operations in phases by January 1998, at a designated location near the Yucca Mountain site. The President has expressed his opposition to the peremptory designation of Nevada as a site for an interim storage facility. The national policy, expressed in existing law, is that Nevada should be excluded from consideration as an interim storage site. The selection of any site for an interim storage facility deserves to be based upon objective analysis using substantive criteria. Furthermore, the relative merits of selecting Nevada as an interim storage site would certainly be influenced by the probability that the repository will also go forward. That probability cannot be evaluated at this time.

S. 1271 provides for the continuation of site characterization activities at Yucca Mountain, and for the licensing and development of a repository if the site is found suitable. In the event of constrained funding, however, the licensing, construction, and operation of the repository are designated to have a lower priority than the interim storage facility and its associated transportation system. Because the measure does not change the budgetary treatment of the Nuclear Waste Fund collections, it seems likely that the funding for the repository determination will be constrained as the requirements of an interim storage initiative increase.

PROPOSED COURSE OF ACTION

Our Administration continues to believe that completion of the scientific studies on Yucca Mountain is the most sensible and principled course of action. If Yucca Mountain is found suitable and meets the test of licensing, provision could have been made to take waste at the site in a receiving facility as early as 2004. The fiscal year 1996 budget submitted by the Administration would have supported this approach. However, funding constraints imposed by the fiscal year 1996 appropriations appear to have foreclosed this schedule.

Faced with those realities, we suggest that the following principles be adopted in legislation. These principles will maintain momentum for permanent disposal, address the interim storage issue, and recognize the realities of the budget outlook.

- Provide adequate funding to resolve the major remaining technical uncertainties at Yucca Mountain.
- Revise repository regulatory structure to reflect current circumstances.
- Select any interim storage site based on objective criteria.
- Initiate generic interim storage activities.
- Protect environment, safety and health.

First, we must fund a program at an adequate level so that an informed judgment can be made of the viability of licensing and constructing a deep geologic repository at the Yucca Mountain site. This judgment must include not only a scientific understanding of the geologic setting, but also an engineering design that confirms that technologies are in hand that can carry out the design concepts, and that the cost and capacity of the potential repository will make it a practical undertaking.

Second, we must revise the regulatory structure that guides the licensing of the repository to reflect the policy changes adopted since enactment of the Nuclear Waste Policy Act, the realities of the budgetary constraints on the program, and the experience of the past decade of work. We must eliminate unnecessary or duplicative requirements that make the cost of repository characterization excessive. In doing this, we will focus on the safety considerations of the proposed project itself.

If the repository regulatory basis is updated to accommodate a more focused approach, and a judgment is made in 1998 that the repository is a practical venture, a license application could be submitted to the Nuclear Regulatory Commission by 2006, which could result in emplacement of waste in a repository beginning in 2015. Our former targets for a license application in 2001 and emplacement in 2010 are nullified by the severe budget reductions in the FY 1996 appropriation.

Third, the selection of any site for an interim storage facility should be based upon the weighing of objective criteria, including cost and social impacts; the technical evaluation of the risk to public and worker health, safety, the environment; and the views of local stakeholders. The consideration of Nevada as a candidate site, or the determination of the need to consider other candidate sites, should await the results of the 1998 viability assessment of the repository.

Fourth, consistent with the approach adopted by the Congress in FY 1996, the program should be authorized and directed to initiate generic work toward early waste acceptance. This work would include requests for proposals for procurements to underwrite the investment necessary for the private sector to develop the capability to carry out the transportation effort that will be required for spent fuel removal from reactor sites nationwide. It could also include the non site-specific engineering work, such as the Topical Safety Analysis Report, needed to begin the licensing process with the Nuclear Regulatory Commission. Considerable progress could be made on these two long lead time requirements in the period prior to the designation of an interim storage site.

And most importantly, any legislation to amend the Nuclear Waste Policy Act must retain the standards and procedures required to assure public and worker health and safety and environmental protection.

STATUS OF THE HIGH-LEVEL WASTE PROGRAM

Program accomplishments

The program accomplishments over the past year represent significant progress toward the requirements of the proposal I have outlined.

In December 1994, we adopted a new program approach that was based on our own critical examination of the civilian radioactive waste management program and on direction and recommendations from the program's oversight, regulatory, and technical review organizations; and feedback from concerned program stakeholders. As an expression of support for the new approach, the Congress agreed to a 40 percent increase in fiscal year 1995 funding to initiate it. Almost all of that increase was allocated to the Yucca Mountain Project.

Since that time we have made real, measurable progress towards geologic disposal. We utilized the entire increase, completing the fiscal year with very little carry-over and with accomplishments that often exceeded our targets. Our tunnel boring machine is two miles into the Yucca Mountain site and well ahead of our planned schedules. We recently completed the turn into the proposed repository formation and have completed four scientific testing alcoves. These alcoves provide our scientists with direct access to the geologic features that must be evaluated to determine the site's suitability for geologic disposal.

With adequate funding, we were scheduled in 1998 to make a determination of the suitability of the Yucca Mountain site for geologic disposal and in 2001, if the site were found suitable, to submit a license application to the Nuclear Regulatory Commission for an authorization to construct the repository. Waste emplacement would have commenced in 2010.

The FY 1996 Energy and Water Development Appropriation Act

The FY 1996 appropriation for the civilian radioactive waste management program is far short of what we had requested. The Energy and Water Development Appropriation Act of 1996 provides a total of $400 million for the program. However, $85 million of the total appropriation has been reserved to be used for an interim storage facility and would become available for expenditure only upon enactment of new statutory authority. Funds to implement the current program are, therefore, effectively reduced to $315 million. The consequence is a 50% cut in the FY 1996 request for the program ($630 million) and a 40% reduction from the FY 1995 level of expenditure ($522 million). Based on Congressional guidance, we have allocated $250 million of the available funds to preserve our core scientific work at the Yucca Mountain site.

Prior to the recent appropriation action, we were on track to meet our major milestones, culminating in emplacement of spent nuclear fuel in a repository by 2010. That program approach is no longer feasible. The $315 million appropriation this year does not support the program approach.

As directed by the Congress in the Conference report on the FY 1996 Energy and Water Development Appropriation Act, we plan to address the major unresolved technical questions regarding the conceptual design of the repository and its expected performance in a geologic setting. Consistent with the guidance, we have defined a new deliverable for the Yucca Mountain Project which would take the form of a "Viability Assessment." This package of deliverables, included in a report to the Congress, would provide the requisite information for a determination in the 1998–1999 time frame of the viability of continuing with the licensing of a geologic repository at the Yucca Mountain site. This work will resolve the critical current uncertainties. With appropriate policy and funding support, we can regain a license application target, but at a later date.

CONCLUDING REMARKS

In closing, I urge you to consider our principles as you mark-up S. 1271. My staff and I are prepared to work with you as your deliberations proceed.

I thank you for the opportunity to appear before you today.

The CHAIRMAN. Thank you very much, Madam Secretary.

Dr. Dreyfus, do you have any statement?

Dr. DREYFUS. No, sir.

The CHAIRMAN. Let me refer, Madam Secretary, to a series of letters that were initiated to the administration. I am sure you recall the correspondence. Last April, I wrote a letter to President Clinton as the new chairman of this committee indicating that one of my top priorities was to help meet the challenge facing the Nation with regard to meeting its obligation for managing spent nuclear fuel. That letter was a consequence of my understanding that the Department had announced that it would not be able to meet its obligation to begin accepting nuclear waste in 1998.

I acknowledged the effort that you and your office were making, as far as addressing this as a top priority. But while you acknowledged that there were national security interests involved, your letter at that time said that you could not support any current legislation to fix the problem at that time. I went on to say in the letter to the President that, if you cannot support current legislative proposals at this time, members of the committee would like to know how and when you plan to offer an alternative proposal. And I went on to say that we had concern about the environmental security implications of failing to reach a solution in the not-too-distant future.

So I concluded by saying to the President that we would urge you respectfully to either support the concept in several current legislative proposals or offer a plan of your own. We have had hearings and so forth and so on. A copy of that letter was sent to you.

In August I wrote again to the President, referring to the letter of April, and my awareness that a letter had been communicated to Senator Bryant from the President stating that the President could not support any spent-fuel management legislation currently before Congress at that time. I raised questions relative to, again, if you cannot support the legislation, what can you support? If you will not support legislation now, when might you support it? If all the comprehensive spent fuel management legislation before the Congress is unacceptable, will you provide us with draft legislation that is acceptable?

And I guess our next correspondence was from Alice Rivlin on August 18, addressed to the chairman, thanking and acknowledging my letters to the President concerning the nuclear waste program. And it goes on to state: As you know, the administration is devoting its full effort to complete the site characterization and other technical aspects of the permanent repository. With respect to proposals that would create an interim storage facility at Yucca Mountain, it states the administration is conducting an internal policy review, as they do with all legislation pending in Congress. The Office of Management and Budget is leading this review in its usual role. The Department of Energy is certainly involved. Other agencies are participating, and we expect to be in a position to communicate an administration policy recommended to me by the time we return from the Labor Day recess. We apologize for the delay.

And that is the last correspondence we have had to date. As you know, Labor Day has come and gone, and the committee is proceeding, and we are still looking for specific direction from the administration. And as one who is sensitive to the concerns of the Nevada delegation and aware of the political realities associated with this issue, it seems to me that the administration has taken a position of not facing up to the reality that it, too, has an obligation, and is either putting it off for the convenience of the political process or awaiting to evaluate, perhaps, a national poll on the issue, but clearly has not met its obligations to this committee.

I am not taking you to task personally, because I happen to be satisfied that a lot of these policies originate, as they should, in the White House, and sometimes you have to be a messenger. But nevertheless, I wanted to convey my extreme disappointment that the administration has not faced its responsibility.

I really do not care to get into the merits of at what phase the responsibility should be addressed and how. But they simply have not met their responsibility in coming forthrightly to this committee with some positive conclusions, but simply have said we are not for this or we are not for that or we are still evaluating it, we are going to need more time, and we will get back to you. Clearly, they have not gotten back to us. Clearly, it is not a priority item, and it should be.

So that is my only statement and question, but I would appreciate your response. And if the administration has a position relative to the communication of last fall that was to be in my hands by Labor Day, why, my office is open any time, or you can slip it under the door at night.

Secretary O'LEARY. Mr. Chairman, I would point out to you, with a great deal of pride, that I do represent the administration. And as you well point out, I am certainly not responsible for answering the mail of others, but I believe this testimony today, filed with the committee on yesterday, is, in fact, a response from our administration.

I probably will get into a great deal of difficulty pointing out to you that my opinion is that you do not like our answer because we have simply said let us complete the scientific evaluation now undergoing, which I know you understand we have increased at a very rapid pace. And our quarrel here is about whether today we make a determination or we make a determination when we have enough facts in hand, as the Congress earlier directed us to, to understand the consequence of the decision that we make. And that has been a very definitive answer coming from our administration for some time.

That we have not formalized it to respond to your letter, I will take care of this afternoon.

The CHAIRMAN. Well, as a followup, maybe a direct question regarding the administration's extent of support for a repository of some kind, either permanent or temporary, at the current site of Yucca Mountain, does the administration support that, or not?

Secretary O'LEARY. The administration does not now support the effort of this committee and others to site an interim facility at Nevada before we have completed the characterization work that we are doing under direction of the Congress. We have been extremely firm on that, Mr. Chairman.

The CHAIRMAN. But you do support continued funding. We have already expended some $4.5 billion and, I do not know, another 3 or 4, will be necessary to bring it to a stage where it can be found either suitable or licensed?

Secretary O'LEARY. Absolutely, sir. And as the Congress has directed, we have reduced the burn rate on the expenditures on characterization somewhat dramatically. The burn rate at the end of this year will be $315 million, reducing to an average of something very close to 4 by the end of fiscal year 1996, and at that time, or a few months beyond that, we should have the answer in hand.

The CHAIRMAN. Well, assuming characterization is completed and licensing is completed and we have expended $7 to $8 billion, is it likely that the position of the administration will be to support the Nevada site at Yucca as a repository for nuclear waste?

Secretary O'LEARY. The chairman understands that as Secretary of Energy I have the responsibility to make a report to the President with respect to the characterization. If it turns out to be positive, then the course we would take would be to move forward with Yucca Flats. I have no answer to that question at the moment, and I do not think it makes sense for me to answer in the hypothetical. I know you would like that, but I do not think it serves any purpose.

The CHAIRMAN. Well, one of the problems some of the appropriators have is whether to continue to appropriate, recognizing that there is a chance, and a reasonable chance, that neither suitability nor licensing will be obtained, and we will have expended $8 billion and will still be addressed with the dilemma of where to have a repository.

Secretary O'LEARY. Mr. Chairman, as one who has had the benefit, and sometimes the pain, of making the decision and advising to take the next step to get the answer, I will tell you that in 2 years and some months we will have the answer, and it seems to me a very valid course to take to complete this action at a much reduced cost, at least on an annual basis. And that is what we are arguing to be able to continue to do.

The CHAIRMAN. I would hope that in your communication to the committee you can persuade the administration to be a little more forthright. That if indeed Yucca meets the criteria associated with suitability and licensing, that the administration will support it as a permanent repository or temporary or some kind, because it is very, very difficult to continue to support funding when there is such an effort, if you will, by the administration to disconnect itself from what taxpayers' funding is going for. The reality is that a permanent repository has to be somewhere, and clearly this one is in Nevada. But it seems like the administration wants to distance itself from identifying with that reality. So with that observation I will turn to my colleague from Louisiana, who owns all the salt mines in the country.

[Laughter.]

Senator JOHNSTON. Thank you, Mr. Chairman.

Madam Secretary, you bravely defended the administration in your questions there. You need not do that again, in my opinion. You are sort of just taking orders on this, unless you want to defend this policy as your own, and I do not believe you should. You should not want to. But in any event, if this is the policy I will want to ask you some questions about it.

Secretary O'LEARY. Senator Johnston, I have to tell you, I know you want to provide me with a graceful exit, I do associate myself with my administration's position.

Senator JOHNSTON. You said that before. I just am not convinced of that.

[Laughter.]

Senator JOHNSTON. On page 7 you said: "if the repository regulatory basis is updated to accommodate the more focused approach and a judgment is made in 1998 that the repository is a practical venture. . . ." Now, the judgment that the repository is a practical venture, is that, to you, suitability?

Secretary O'LEARY. That would have been suitability before the last budget actions. It is something less than the suitability review and assessment that we took direction from the Congress to undertake.

Senator JOHNSTON. Now, you said in a statement along with Alice Rivlin to the President, a letter, that you could make, in fact, a suitability decision, as I read the letter, by 1998. And this is today's statement, and today's statement says that if a judgment is made in 1998 that the repository is a practical venture. My question is, is today's statement still a valid statement?

Secretary O'LEARY. It is, but I need to clarify. The language we were using, Alice and I, in the letter that came to the chairman earlier was pre-fiscal year 1996 budget. And all of the assessments and technical work that Dr. Dreyfus and his staff has proposed to undertake to complete the suitability assessment have now been reduced dramatically, occasioned by the fact that the budget has been reduced.

Senator JOHNSTON. Well, I am saying this is today's statement. Is this still an operative statement?

Secretary O'LEARY. That is correct, and the distinction I am trying to make is yes, we will do the work, we will meet the timetable, it will be fewer studies, fewer bore holes than we had anticipated.

Senator JOHNSTON. Well, when you say judgment is made that the repository is a practical venture, is that suitability?

Secretary O'LEARY. And what I am trying to do now is make this very fine distinction between what we had earlier determined to be suitability to tell you we can come to conclusions. We will not have the wealth of data and information and studies we had planned to have, occasioned by this reduction in budget. Therefore, we are calling this assessment something less than suitability, and the phrase we have come up with is feasibility.

Senator JOHNSTON. In your letter to the President you said that in January 1998 you can make a judgment as to the suitability of the repository, and at that time if it was suitable then you could commence, at that time, with the interim repository. Is this the same judgment? Can you still make the judgment in 1998 to trigger the construction of the interim facility?

Secretary O'LEARY. The Department can make the judgment. Our process then would be to make that recommendation to the President, and he would then take an action, and that is now required in the statute. So yes, we can undertake and discharge our responsibility.

Senator JOHNSTON. In 1998?

Secretary O'LEARY. That is correct.

Senator JOHNSTON. All right. Now, if you will look over on page 6 of your statement, you say "we must fund a program at an adequate level so that an informed judgment can be made of the viability of licensing and constructing"—licensing and constructing—"a deep geological repository at the Yucca Mountain site." This must include scientific understanding, and also an engineering design that confirms that technologies are in hand that carry out the design concepts, and that the cost and capacity of the potential repository will make it a practical undertaking.

Now, is this the same thing as you were saying on page 7 that you could do by 1998?

Secretary O'LEARY. Exactly.

Senator JOHNSTON. You are saying that in order to make your judgment in 1998 you would have to have a design and you would have to know about the viability of licensing?

Secretary O'LEARY. I am saying we would have to have a design. The design would be somewhat different and less tested than we had earlier planned, but yes, we would have a design, and I believe we would know. I know that we would understand the viability of licensing. But I am very clear that we would not have had the benefit of all of the information we would have liked to have given to the NRC, nor would we have had the exchange between and among the staff at the NRC on the details of the data we would have been developing.

Senator JOHNSTON. Now, EPA is developing some standards. Do you need those standards before you design the repository?

Secretary O'LEARY. I believe that it would be very useful to understand the standards against which we would be building the repository, absolutely.

Senator JOHNSTON. Dr. Dreyfus, you just wrote the EPA after having received the report of the National Academy of Sciences in which you stated how important it was that they have viable standards. And part of the sense of your letter which I understood was you really cannot design until you get those standards. Did I read that incorrectly?

Dr. DREYFUS. As far as our current strategy for design is, our waste isolation strategy assumes we are going to have to minimize the peak exposure, which is at a very long point in time. So we are basically designing against the anticipation that the standard is going to look like the Academy suggestion. What we are asking EPA to do is in writing the standard itself to dispose of some of the more problematic judgmental aspects of that thing so that we are not in a position of predicting what will happen where one can only assume what will happen.

Senator JOHNSTON. With today's law, can you really design without knowing what those standards are?

Dr. DREYFUS. We are designing against worst-case.

The CHAIRMAN. Could you repeat your response, please?

Dr. DREYFUS. I am designing against a worst-case outcome of the standards.

Senator JOHNSTON. And how long would it take you to get that design done?

Dr. DREYFUS. Under the current budget profile that we have, we are working on a waste package and the underground safety considerations and things like travel time to the environment, so we will pursue all of those things that are critical to a total system performance assessment. What we are not any longer doing is bringing along to the same degree things like the surface facility design and ancillary work that we pretty much know how to do, but we are not going to bring the design up to the same degree of perfection. So this is very focused.

Senator JOHNSTON. I am not trying to get you in trouble with your Secretary, but can you design by 1998, really?

Dr. DREYFUS. I can determine by 1998 whether this thing is technically feasible. I will not have a design to submit to the commission because the nonessential parts of it will be left conceptual, and we will concentrate on things that are critical to the waste isolation strategy.

Senator JOHNSTON. So when the statement here says that the judgment must include an engineering design that confirms that technologies are in hand to carry out the design concepts, you are not going to have that in 1998.

Dr. DREYFUS. We are going to have that for all of those areas where technology are in question. I am not going to have it for road, surface buildings, handling facilities on the surface, a lot of stuff that we know pretty much how to build. But we will leave that design in the conceptual situation it is in. We will work on the waste package and the placement and those aspects that are critical to the waste containment.

Senator JOHNSTON. So by 1998 you can know, Madam Secretary, whether this is a suitable facility?

Secretary O'LEARY. Know whether it is suitable? As I understand it, I clearly understand the concept of a reference-cased design which permits you to make all of the assumptions that are ready to be made and to do a design.

Senator JOHNSTON. Let me be clear. When I say suitable I mean suitable to trigger the selection of Yucca Mountain as the interim storage facility.

Secretary O'LEARY. Absolutely, Senator. And that is what I think we have all been committed to have done.

Senator JOHNSTON. All right. So if we pass legislation here that requires that DOE make its suitability determination by January 1, 1998, and I think we have to define suitability, and the way I defined it in the appropriation bill was that you can design and construct a facility of useful size with a high probability of being able to construct and design one that could be licensed. You could meet that test?

Secretary O'LEARY. The answer is yes.

Dr. DREYFUS. Well, I think you just took away 1998. Did you say January 1, 1998? Well, I need more time than that in 1998.

Senator JOHNSTON. I did say January 1, 1998. How much time would you need?

Dr. DREYFUS. Well, I have not got the thing that boiled down.

Senator JOHNSTON. How about July 31, 1998?

Dr. DREYFUS. That is only 2 years. I need 3 years.

Senator JOHNSTON. You will be past the Ghost Dance fault very soon, will you not?

Dr. DREYFUS. Well, it is the design work and the modeling work. The tunnel will go along quite well. I have got a heated test going in I need some data from. There are timelines in here. We need to have the 1998 work in the fiscal year, October.

Senator JOHNSTON. All right. Well, we can work with you on a date within 1998.

Dr. DREYFUS. Give me till October.

Senator JOHNSTON. To say you must determine suitability by that point.

Dr. DREYFUS. Yes.

Senator JOHNSTON. And you will be able to meet that?

Dr. DREYFUS. Yes, sir.

Senator JOHNSTON. All right. Now, in order to begin construction—first of all, Madam Secretary, what do you figure a year of delay will cost the utilities at the end of this and having an interim storage facility? Would you agree with me it is something over a half-billion dollars a year?

Secretary O'LEARY. I agree with you that that is the testimony that I have read and facts that I have reviewed. Yes, I agree with you.

Senator JOHNSTON. So it is worthwhile to try to avoid this delay, and therefore we ought to be ready to go in constructing the interim facility as soon as we have determined the suitability of the site, should we not?

Secretary O'LEARY. Absolutely. We have always agreed on that point.

Senator JOHNSTON. Now, according to Dr. Dreyfus' interim storage summary schedule, which he submitted to me at my request with the appropriation language which I submitted, we asked how long would it take to be able to store waste, to commence the various critical path items in order to store waste in an interim facility. And Dr. Dreyfus indicated that you could complete construction by January 1, 2000, provided you identified your site by January 1, 1996. Well, we will not make that, so we will have to slip that.

But if you began construction on January 1, 1998, you could in fact begin to store waste by January 1, 2000. But in order to commence that construction by January 1, 1998, you had to begin Phase I design, you had to begin site assessment, you had to begin your environmental work, you had to begin your procurement of casks, equipment, and services and utility interactions, you needed to do all that this year, is that correct?

Dr. DREYFUS. You need 4 years from the time you fire the starting gun, and I must point out that is assuming that you are doing a licensing approach such as is in the Craig bill or the Upton bill. Under existing law, we are back where we were the first time we discussed this.

Senator JOHNSTON. I understand. We would have to change the law. So really what I am getting at is if we want to avoid at least 2 years of delay, then it seems to me that we ought to begin all of this work on the interim storage facility other than the construction, we ought to begin that now. Otherwise, we lose 2 years on the end of the schedule, and as I understood your letter, Madam Secretary, along with Alice Rivlin, to the President, you said we should not build an interim storage facility at Yucca Mountain unless it is the site of the repository, of the permanent repository.

So does it not make sense to say that we ought now to pass legislation that puts in process these things like site-assessment design, negotiations on intermodal site, requests for proposals from the National Transportation capability, all of those things that Dr. Dreyfus had identified that need to be done in 1996 and 1997, short of construction, should we not begin those now, if we are going to begin construction upon determining that the site is suitable?

Secretary O'LEARY. Senator, you will see in my written testimony that I am clear that the generic activities we can undertake to help us move speedily once we know if the interim facility will be located in Nevada ought to be done. But I do not believe that we should trigger those activities which presume the decision that is yet to be made.

Senator JOHNSTON. Well, suppose the legislation says that if the site is not suitable, the repository site is found by you to be not suitable after these extensive studies, that then you may not build or build either the interim or the repository, and you must select an alternative site. Would that not satisfy the requirements of your letter?

Secretary O'LEARY. I do not believe so, and I make this statement on the basis of my experience. Normally, a thing begun, if the cost is de minimus, continues whether it is appropriated or not. And I once again would state to you I believe those generic activities ought to be started. We are prepared to do that.

Senator JOHNSTON. Do you think we ought to delay this 2 years because you cannot trust yourself to terminate if you find it not suitable?

Secretary O'LEARY. I trust myself. I do not necessarily trust everyone else on this issue.

Senator JOHNSTON. Well, Madam Secretary, forgive me, I read your letter very carefully, and it seems to me you are adding an additional element now. In addition to saying you should not construct it at a place where you are not going to have the repository, you are saying you should not even think about it. You should not even study it. You should not do anything. Wait another 2 years. Wait, wait, wait, wait.

Secretary O'LEARY. You misunderstand both my letter and you certainly misunderstand the testimony submitted. I am saying everything that we can do that would be generic and would help us move along ought be done.

Senator JOHNSTON. Well, yes, but doing those generic things does not help your critical path here. You have got 2 years of site-specific work, like an EIS, like site characterization, site assessment, and all that, that is short of construction, but which is a necessary predicate to begin construction. And you can save at least 2 years or more if you will begin that work now.

Secretary O'LEARY. I understand your point, and I agree with the facts that you point out. With respect to saving time, I would point out to you that I believe it is very important that we build both confidence and trust as we move along this process and to trigger early before a decision I do not believe does that.

Senator JOHNSTON. All right. Let me just ask you one more question. I am sorry to take so long. On page 6 you say we should revise the regulatory structure. What does that mean?

Secretary O'LEARY. I am going to let Dr. Dreyfus respond to that, because that is definitely his section.

Dr. DREYFUS. In the last year or so we have had a lot of interaction with the Nuclear Regulatory Commission, and it has become reasonable clear to me that the regulations we are operating under, most of which were written pursuant to the 1982 Act, involved a lot of activity that is largely associated with selecting among alternative sites. We are not now doing that. We ought to come to a proposal for a specific repository concept at Yucca Mountain and make a safety case for that proposal. That is substantially different from the concepts in our regulations, which I am prepared to revise, and the Commission's regulations, and I think that we could much enhance our ability to focus this decision and reduce the cost if we set about revising those regulations accordingly.

Senator JOHNSTON. Do you need any legislation to trigger that revision?

Dr. DREYFUS. Well, my judgment is that we can do a lot without legislation, but that in revising regs it is helpful to have new guidance. Now, we have some. We have never recognized the 1987 Act in the regulations, and we certainly have not recognized the most recent Appropriation Act, but more direct guidance would probably be helpful.

Senator JOHNSTON. Well, I hope you will submit that legislation to us, if we can do that.

Thank you, Mr. Chairman, for your forbearance.

The CHAIRMAN. Thank you very much.

Senator Craig—excuse me, Senator Kyl was here first.

Senator KYL. Mr. Chairman, I am happy to have Senator Craig proceed, as long as I have a couple of minutes before I have to go preside.

The CHAIRMAN. All right.

Senator CRAIG. In deference to time, I will be reasonably brief. Because I think both what you have done, Mr. Chairman, and what Senator Johnston has done, and the line of questioning they pursued with the Secretary is clearly the points that I wanted to make.

Madam Secretary, you just made the point, when asked by Senator Johnston, are you willing to move ahead in reasonable, responsible, and expeditable ways, based on site characterization that you say will be completed in 1998, and let me tell you, if it is, you win the prize, because nothing has been met to date on that site yet in any timetable projected. No, I am not accusing you.

Madam Secretary, I have one reaction to what you have just said to both of these gentlemen. In the context that you are the administration speaking at this moment, you are stalling. I have heard a very sophisticated first-class, articulate stall. That is what you are doing. And you are stalling for political reasons, and that is tragically disappointing.

Are you prepared in 1998, under the 1985 Act, or 1982 Act, to stonewall the utilities from your legal responsibility to start accepting their waste? I mean, is that the game that is out there now, and does that fit the pattern of a variety of people inside this administration who are boldly antinuclear who say if we can just stall it long enough we will kill them all and they will go away?

Now, I have asked you a series of questions, but I do believe in the first instance your first answer and your entire statement is a stall.

Secretary O'LEARY. Senator Craig, I believe that this team of people in the Department of Energy has done anything other than stall on the requirements that they had.

Senator CRAIG. No, no, no. Not on the schedule that Dan Dreyfus is on, on the fact that you will not support action by this Congress and become an active part of preparing for an interim facility if the standards and the site characterization are met. And you just said no. And so you will not have a site online for receiving interim facility storage down the road. You are stalling, you will waste 2 to 3 to 5 years at best, even on that characterization. That is what I drew from that dialogue that just concluded.

Secretary O'LEARY. Well, I beg to differ with you.

Senator CRAIG. Please do. I need satisfaction.

Secretary O'LEARY. Well, I hope that I will be able to provide it. Where you and I disagree is on the timeline to move along that course. And you may choose to call that stalling. I call that responsible appropriate action. We are quarreling, and you and I are disagreeing, on 3 to 2 years. But for goodness sake, Senator, I am so glad that others are not here, but much more time has been lost than that, and we are trying to build a progress that creates some confidence in the action that we take, and that is something that this program has lacked for decades.

So you may be unhappy with me, but I am anxious to work with you, and I believe the way we proceed is with due responsibility to timetable, but also with due responsibility to building the scientific base to get the trust we need to go forward. And I, certainly more than any, represent myself as being very supportive of the nuclear industry. It provides 21 percent of the power that generates this economy.

Senator CRAIG. Well, it may provide a lot less if we do not act immediately on the timeline that is obviously necessary and required just to move to compliance, making the assumption that you meet your 1998 timeline. And you are saying no, let us wait till 1998.

Secretary O'LEARY. And I would also point out to you that this Secretary and this Director have met their timelines.

Senator CRAIG. In 1987, if we had continued the fight over regional repositories, as most wanted us to do, certainly the antinuclear community did, if Congress had not acted then we would not be here today. We had to force the issue. And my guess is we have to force it again.

Your administration and you, and I am glad you did, are signing contracts or making agreements, court-binding legal agreements with States like Idaho to do certain things in a timely fashion, and yet it appears in some instances that actions do not fit those obligations. How do we get there?

Secretary O'LEARY. Senator, I think our actions do fit those obligations. We have a legally binding obligation with the State of Colorado, with the State of New York, and now with the State of Idaho for dates certain that I believe that we can meet. And within the timetable we have outlined in my testimony and the letter that Alice Rivlin and I have sent to you, so will we meet our commitments to Governor Batt and the Governors of the States of New York and Colorado.

I am very clear on those responsibilities, and have been since I arrived in this job. And we can continue to quarrel on the 2 to 3 years upon which we differ. What I would like to see is not only legislation, but I would like to see a facility built.

Senator CRAIG. Then let me ask this question of you. If you reject the legislation that you came to testify on today, and you do, and if Senator Johnston and I can sit down together and work out our differences in cooperation with you, and part of that agreement is to move forward with design and the necessary processes to expedite and facilitate action toward the construction of an interim facility pending a 1998 site characterization that you can then report to the President is satisfactory for the purpose of building a deep geologic repository at Yucca Mountain, and that that agreement, that legislation, passed by Congress, laid on the President's desk, suggests that we start acting now in the expenditure of the $85 million necessary to get the designs in place, to get everything ready to go in advance, so that if your site characterizations and Dans works, would you agree to that?

Secretary O'LEARY. I will agree to all that has to do with generic preparation, not with specific preparation directed at Yucca Mountain or nearby facilities.

Senator CRAIG. Because that is the political side that you cannot agree with.

Secretary O'LEARY. It is the reality that it does not, in my view and in the view of our administration, seem responsible to begin this work without the answer. And I am clear on generic work.

Senator CRAIG. We cannot spend $170 million, or a few dollars more, in light of the billions that you are now spending to get to that 1998 deadline? That is not responsible?

Secretary O'LEARY. Spending approximately $115 million this year, less next year, and that is the point on which we differ, and you and I will differ on this point for as long as I am testifying today.

Senator CRAIG. Well, this Senator, and I think you have heard the Senator from Louisiana, are committed to trying to resolve this issue. We want to work with you. We want to expedite this so you do not get the heck sued out of you, and you are going to spend millions and millions of dollars in court for no reason. But again, it is a cheap game. It gets paid, and it is ratepayer money and it is taxpayer money and it ought not have to happen.

Now, it was not your fault, but you are now in charge. And what is your responsibility, Madam Secretary, in my opinion, is to not only demonstrate that you can meet your 1998 deadline, but to show all other actions that expedite, if characterization is affirmative, the action of bringing about an interim facility and a permanent repository. And you are just telling us no, we can only generically move along until 1998. That is unacceptable.

Secretary O'LEARY. Then I take your declaration that my testimony is unacceptable.

I would like to say this about the lawsuits. The lawsuits that have been filed against the Department of Energy now are directed for both parties, the defendant and the U.S. Government, the Department of Energy, to mediate to conclusion. The Department is happy to take this course to avoid the description that you have just laid out where we are sued and the taxpayers pay unnecessarily. We are very open to this process.

Senator CRAIG. Thank you.

Thank you, Mr. Chairman.

The CHAIRMAN. Senator Kyl.

Senator KYL. First of all, may I just follow up on something. I think, Dr. Dreyfus, you said that you would be willing to submit your recommendations for regulatory changes that would expedite the procedures, am I correct on that? And, Madam Secretary, would you be willing to do that just as soon as you possibly can, so that to the extent that we can assist in identifying what has to be done legislatively we can do that very quickly?

Secretary O'LEARY. Absolutely, Senator Kyl. And I have attempted to make it clear all morning that I stand ready to work with this committee as well as its chairman and the proposers of this legislation to try to come to some conclusion that reaches the goals that we have all espoused.

Senator KYL. Rather than waiting for us, I would appreciate it if you would submit those just as quickly as you can.

It seems to me, Mr. Chairman, that we are forgetting something, and that is the amount of material that is being stored onsite around this country. I think it is 75 different sites where commercial production is engaged in there are something like 30,000 metric tons of this material, and that does not take into account the defense material, the Department of Energy material, or the Navy material. And it seems to me that one of two things are true, either there is a big problem with this material, or there is not. We have spent $2.7 billion for site characterization, presumably for some reason.

There is some reason we have spent all of that money and taken all this time, and there is some reason that we have to take more time. And presumably that has to do with just absolutely making sure that we store this stuff safely at the same time that for years now and for years to come this material is sitting in ponds and in certain cases some dry storage at 75 different sites around this country, at the commercial sites, with very little concern apparently being given to the fact that if there is a danger it already exists.

It seems to me that we are being very hypocritical here. We are willing to let something sit there which obviously poses a problem and a potential danger for years while we study how we can safely store it in the far, far, far distant future. There is something drastically wrong here. If we have to spend all of that money and take all of this time finding a safe site, then maybe we ought to be more concerned about the interim storage of what we already have. And it seems to me that we are making perfection the energy of the good here, and that what the Department ought to be doing is finding a way to very quickly provide interim storage for the material that is piling up around this country.

At the Palo Verde site, the largest nuclear generating facility in this country in my home State of Arizona, and Richard Snell, who is the chairman of the board of the parent corporation there is here today to testify, he is going to have to respond to the people in Arizona who have spent over $150 million already, and what do they have to show for it? Nothing. And that site is rapidly accumulating the products which are going to have to be stored here pretty soon.

How do we respond to our constituents who say what have you gotten for that $2.7 billion in site characterization, and if it is so important to be concerned about the safety, why are you not concerned about the safety of what is already out there? And if it really is not that big of a problem, then why do you have to spend so much money and time in developing this new site?

On page 4 of your testimony, Madam Secretary, at the bottom of the page you said the relative merits of selecting Nevada as an interim storage site would certainly be influenced by the probability that the repository will also go forward, meaning the permanent repository. Why would that be the case?

Secretary O'LEARY. Because then one begins to plan certain that a receiving facility or an interim facility to take waste will be just that, an interim facility, and we need not last longer than the already certified life for these casks now at site, at utility, and waste in pool. And I want to be clear to state that under the competent hearing of the NRC, both in the early 80's and 1987, these dry casks have been certified to be safe for approximately 100 years, and there has never been a question in the recent decade that storage of spent fuel in pools appropriately done, as is the case with industry, is anything other than safe. And the NRC refuses each time it does an inspection at site.

Senator KYL. Well then, what is the problem with sending it to this interim storage site at Yucca Mountain?

Secretary O'LEARY. The problem is that the requirement, sir, forces us, as it should appropriately, to look at a much longer term for storage that goes well beyond the 100 years that is now permitted within the commercial industry, and the responsibility to understand that, quite frankly, has only been undertaken with a great deal of vigor in the past 5 to 6 years.

So the answers are not there that are now required by the Act. We are attempting to get that work done. When we came into office we said we could have those answers in 1998, and we will have them in 1998.

Senator KYL. Well, the very next sentence of your statement says that probability cannot be evaluated at this time; in other words, the probability that the repository will go forward. So if you say it cannot be evaluated at this time, really all we have said so far is virtually meaningless. We are aiming toward a data and a decision which you have said cannot be evaluated at this time. We may find that that site is not suitable, then what?

Secretary O'LEARY. No, what the import of the sentence, as inelegantly drafted as it is, indicates that today we cannot tell you with certainty that the site is suitable. We will be, at the time we complete our work, and we have indicated that we will do that on time.

Senator KYL. Well, what is the probability of the finding of suitability at this point, then?

Secretary O'LEARY. We have had discussion on the record on this question at our last hearing, and I would hate to have to be the determiner, but Dr. Dreyfus wants to do it, so let him do it.

Dr. DREYFUS. It is better than it was at the last hearing. We have had a very eventful year. We have got a tunnel into the repository formation, and tunneling has confirmed most of the surface work that we have done and most of the modeling assumptions that we have made. At this point, there is no reason to suppose that you cannot use the site. The next 6 or 7 months are going to, again, present us with a whole lot of confirmatory information we did not have before.

I believe that there is a very, very high probability that we can design this repository and that it will perform against any reasonable standards. There remains a rather Byzantine regulatory showing to be made, and that, I think, can be done, but it is a ratio of time and cost against probability of success.

With enough money, we can make the showing. But whether the Congress will give us the money in time to make that showing is something to reckon with. So I always place the probability of regulatory success below the probability that we have got a site.

Senator KYL. And that is a self-imposed limitation in the sense that the Congress——

Dr. DREYFUS. It is if you look at it from the point of view of the Nation at large. From my point of view, it is not self-imposed, and I simply have to make the case, and I can evaluate my chances of doing it, and with enough money I can probably do it at the same probability of success as a repository. But nothing that has happened in the last year comforts me that I will be able to do the amount of work that the current rates require.

Senator KYL. And that is why we would like to make the revision, so that we are not spending unnecessary money or time chasing an nth degree which is really not necessary to make a practical decision that the Secretary spoke of earlier.

Given the fact that you have said there is no reason to believe that we cannot use the site, and that your other phrase was that there is a very high probability, I go back to Senator Craig's question: Why should we not start now to begin the process of preparing this interim site, since there appears to be very little likelihood that that site will not be able to be used in conjunction with the permanent site? It just does not seem to me that there is very much to lose by beginning now and preparing this as the interim site.

Secretary O'LEARY. Senator, I have had the privilege and the pleasure of living in Minnesota, in Tennessee, Virginia, and New Jersey. All of those States were originally listed in the winnowed-down list of I believe the seven or nine States that may have been the target for the characterization or the examination that now Yucca Mountain is undergoing. As I think back to the times when I lived in those States and the anguish that the people of those States had that they might be the site, I believe one of the reasons, as I have indicated before, that we must proceed step-by-step so as to build the confidence on the part of the public generally and now the citizens of Nevada specifically, that we know the course we take, and we have taken all appropriate care to undertake each and every investigation required by the statute. It is for that reason that I sincerely believe we need to take the course that our administration outlines.

Senator KYL. Well, the people of Nevada who oppose this are never going to be satisfied by the scientific findings, because they oppose it for other reasons, and one can understand those reasons. They do not care whether the science demonstrates that it is safe, they would just as soon not have it there, and I do not blame them. And those who support it understand that it can provide some economic benefits of the State and that science will prove it to be valid, and therefore not a problem. I understand why we have to make sure that we are doing the right thing scientifically. Nobody denies that. But you could fight two World War II's in the time that it has taken us just to figure out whether we are going to dig this tunnel or not.

I said self-imposed before from the standpoint of the Congress. We impose the various regulations and we can change that. And you have made a construction that you can show us some—I do not want to use the word shortcuts, because that suggests maybe that we are not adequately looking at it, but some ways of doing this that can save us time. Because, as was pointed out earlier, Senator Johnston, I think, pointed out that every year it is a half-a-billion dollars just for the commercial industry, and that does not count all the other costs.

Time is money here, and it just seems to me that we have just reached a decision that we do not want to do anything for 2 years, and therefore we have constructed some reasons why not to do it. When instead, Madam Secretary, with all due respect, you ought to be trying to find some ways to get it done. And given the fact that there is a high probability that Yucca Mountain will work, that there is no reason to believe that we cannot use the site, that there is this cost ever year while we delay it, that there is very little down-side to doing it, that while it is safe to store this material at other sites there are problems in storing it at the other sites, we ought to get on with the process.

Secretary O'LEARY. And I think that you and I agree that that should occur, and where we disagree is how much should be done. We and our administration are in favor of beginning the generic work, and I understand your point.

Senator KYL. Thank you, Mr. Chairman.

The CHAIRMAN. Well, we have reinforced the generic effort.

Senator Grams.

Senator GRAMS. Thank you very much, Mr. Chairman.

Madam Secretary, it is a pleasure again to have you here. I would like to shift gears a little bit. Probably no more comfortable, though, but as you know, I am introducing legislation today that is aimed at trying to privatize the interim storage program, and I would like to talk to you a little bit about some of the serious issues of interim storage. Many utilities such as NSP in my home State of Minnesota are facing very difficult decisions in providing adequate storage space for spent fuel that is being generated by their nuclear plants, and in Minnesota, about 30 percent of our power comes from nuclear energy.

I know, as you are well aware, and we are facing now a deadline of 2002, and as we go through all these scenarios we are not even coming close to that date, so NSP is going to hit a wall, along with many other utilities in this country, much, much quicker than 2002, and with the legislature in Minnesota putting some real tough restrictions on this and with no additional space it could cause the beginning of the shutdown of many of these plants around the country, which for many people is a goal.

But over the past 13 years we have seen the Federal Government collect $11 billion, spend about $5 billion of that, and we are no closer today. And I know you are coming in in a very tough spot, and you have probably made more progress over the last year than has been made in 5, so what we are talking about, though, is the Government's overall failure to deliver on this type of a situation, and I believe in one instance we can take a lesson from the private sector here.

A number of utilities, including NSP, have been working with the Mescalero Apache Tribe of New Mexico in a joint effort to design and to license a privately operated interim spent fuel storage facility. Now, it is less costly, under estimates than the program that the DOE is proposing, but the joint utility effort and the Mescalero Tribe effort alone is not enough, of course, to resolve this issue, and I believe we should let the private sector, and this is why I am proposing this part of it, not the DOE, develop, license, build, and operate this storage.

Given the delays and the problems that we are facing, Madam Secretary, associated with the federally managed Office of Civilian Radio Active Waste Management, could you support the privatization of the program as an option?

Secretary O'LEARY. Last year, Senator Grams, or maybe it was 6 or 7 months ago when I had the opportunity to consider that concept I was very clear to say I believed it was fair that that would be an option if the Department of Energy could not and did not meet its milestones that we were publicly committed to make. I find that, one, we have met our milestones, we are ahead of them by some 40 percent, we are on budget in this program, and it occurs to me again, after some years of experience both in and out of Government, that you are likely to lose more time attempting to privatize the activity, because we will go through, in my experience, the 4 to 5 years of convincing all the people who need to be convinced that this is an appropriate activity.

So I suspect that we could perhaps finish on the course that we are on much more quickly than to implement the proposal to privatize, and that has certainly been my experience with the U.S. Enrichment Corporation.

Senator GRAMS. Would you not agree, though, that the private sector may be able to more efficiently license, build, and operate such a site, and probably much quicker than the DOE? And if so, why not just guarantee the option of this as a dual track? Because I think, as we heard here again this morning, we might hit 1998 and the year 2000 and the year 2002 and still not have the answers that are going to help NSP and other utilities with this problem, and this has been coming for 16 and 18 years, and now we are saying another 2 years or 4 years will not hurt. It will.

Secretary O'LEARY. Every year hurts. We are clear on that. We want to stay on progress.

I heard your opening statement with respect to the Mescalero business plan, and my recollection was that they had indicated that they could deliver in 1995. I made a phone call to check, and that continues to be the case. But I am not prepared here to embrace

23-370 - 96 - 4

an option for our administration. What I will tell you is that in looking at all the options that might be on the table, clearly that is one. And what we must now ask ourselves, and that must be done through a variety of public and open processes, is can you get this idea up and running and producing before Dr. Dreyfus and his team can complete the course that they are now very soundly on? I do not believe you can, but I think it is something that ought to be examined.

Senator GRAMS. Now, I am also aware of your concern over the lack of requirements at the Department relative to NEPA and the licensing of the facility, and of course this proposal, the private facility, would have to receive its operating license from the NRC. That is the correct place for applications, and it is important for public participation, as well, and that the NRC would have to be involved in the licensing, completing of the EIS. So the language in my proposal merely would try to streamline the efforts, eliminates the duplicative consideration by DOE.

So again I am going to ask you, do you not think there could be some efficiencies, especially for the taxpayer, and also to try to ensure that there would be a spot for storage gained by having a private consortium construct and operate such an interim facility, and there are plans on the table, places eager to begin this type of study and everything, then lease this facility to the DOE, in other words take the DOE out of the loop except for working on the lease proposals?

Secretary O'LEARY. I understand the theory of your concept, and what I would propose is that we look at the progress that Dr. Dreyfus is making against the progress that the Mescaleros are making, and let us ask ourselves in another month whether your suggestion meets the bill in terms of timetable. And I am not so certain that it will in terms of cost, but I am willing to be persuaded that it will, because I, too, have been engaged in some privatization, and when it accords and takes care of the rights of all of the workers involved, sometimes it does have a great—it makes a great effort and succeeds in reducing costs.

So I am not turning this away out of hand, but I am saying before we embrace it entirely, let us understand if it accomplishes the goal we both seek.

Senator GRAMS. Madam Secretary, I have one other question I wanted to ask about a small power cooperative known as Dairyland. You might be familiar with this, one of the test models built by the Atomic Energy Commission, and these small models were later sold off to various entities with the presumption that they would be able also to reprocess spent fuel and they would not have to worry about this problem about spent fuel. However, the DOE now is arguing, it is my understanding, that they do not have to accept waste from this facility—primarily Dairyland—even though the DOE and policies changed the rules after they purchased it. Right now, and I am asking for them, all they want to know now is will their waste be accepted by the DOE as well in the future?

Secretary O'LEARY. As I understand it from just this very quickly whispered conversation with my colleague Lake Barrett, at issue is not whether we will but the timetable in which we will. Let me commit to do this, Senator GRAMS. I will go back today, I will understand this issue both with respect to our obligation, which I understand clearly, I suspect the issue may be contract and time and que, but I will get to the bottom of this and I will phone this before I leave my office this evening.

Senator GRAMS. Okay. I appreciate it. Thank you very much.

Mr. Chairman, I do not have any other questions. Thank you.

The CHAIRMAN. I just have one other question, and it has been brought up by other Senators, but I want to make sure I understand. You say, Madam Secretary, that you will not support funding of those things that could be done prior to a specific site being designated?

Secretary O'LEARY. Well, that may be a little close to the mark, Senator. I do hate to repeat this again, because it simply annoys you.

The CHAIRMAN. Well, I asked a question in just a few words to try to keep it rather direct, and please try and respond to that in that dimension.

Secretary O'LEARY. Well, you and I differ, I believe, is on the work to be done on the interim concept. What I am hearing from the suggestion from you and Senators Johnston and Craig and Kyl is that we should do all work specifically related to the site before we understand whether that site can be the permanent repository.

The CHAIRMAN. That is pretty much our contention.

Secretary O'LEARY. And I am saying no. We do the generic work.

The CHAIRMAN. And you are saying no, and I guess I am not satisfied that you are giving an explanation as to why the answer is no, when we are going to be committed to expending funds on this site, because you have to have some site to start the process. So it is kind of the question of the chicken and the egg, in that sense. And I just do not understand, nor do I appreciate, I guess, your explanation as to why you could not proceed as though this were the specific site, since there is so much funding and expertise going into what has been designated as the likely specific site. And since we have dragged this on so much and spent so much money, it would seem to me that you certainly have a consensus among members here that it would be more expeditious to proceed. I do not see the down-side.

Secretary O'LEARY. Mr. Chairman, rather than have me say no for the rest of the morning and you to just say I do not understand, I think it would make just more sense for us to put our staffs together and let us clearly understand from you the steps you want to take, the activities that could be undertaken to expend the $85 million that is now fenced. My major concern would be since I recognize there will not be copious funds of money available that we expend this money to no purpose and have difficulty finding it if we need it in another setting. But let us put our people together to understand where we differ.

The CHAIRMAN. Well, I appreciate you coming halfway, and I am sure that the appropriations members that are on this committee, including Senator Hatfield and Senator Johnston, and occasionally I have Senator Stevens telephone number, would be sensitive, if you will, to those concerns. Your willingness to explore this with the professional staff and the individual members' staff, I think, is meritorious.

My only other question relative to funding and the recognition that we have got the Nuclear Waste Policy Act to adhere to as well as NEPA. This would be to Dr. Dreyfus regarding the funds on hand, the schedule, and I guess more directly, your attempting to proceed as you have to under the existing law and regulatory authority. Do you have any recommendations to this committee that would specifically expedite your responsibility and save the taxpayers money, and perhaps reduce duplication of the regulatory oversight, whether it be NEPA or the Nuclear Waste Policy Act?

Dr. DREYFUS. The only recommendation that I have got is I do believe that there could be a more focused approach to the licensing showings that have to be made with regard to the repository. I think it can be done without changing in any respect the essence of the safety case, and it is unrelated to standards. It is a matter of process and procedure. I do believe that there is a substantial amount of money involved in the difference between having a proposal and making the showing that I think could be saved by changing the regs.

The CHAIRMAN. Would you have any hesitation regarding, and Madam Secretary, would you support, the submission to the committee of that suggestion?

Secretary O'LEARY. Yes. We have already indicated we would be pleased to.

The CHAIRMAN. All right. I have no further questions.

Senator Domenici, we are still on the first panel. You are here. It is your turn.

Senator DOMENICI. I thought you were going to suggest that I not say anything.

[Laughter.]

The CHAIRMAN. I would not make that broad suggestion.

Senator JOHNSTON. That was the hope.

[Laughter.]

The CHAIRMAN. I may give you the gavel here in about 5 seconds.

Senator DOMENICI. Well, I hear you talked so long, Senator Johnston, somebody left.

[Laughter.]

The CHAIRMAN. Senator Wellstone left.

Senator DOMENICI. Well, I said somebody.

[Laughter.]

Senator DOMENICI. I do not intend to do that.

Actually, you know, Senator Johnston and I switched positions. He used to chair the subcommittee that had to put up with all of the excuses from everyone about this program, not limited to you, but including our President and the administration. And now I chair it this year, and obviously some of this testimony from your side has addressed the issue of whether or not the program that we funded has the kind of money in it that will let you do what you want to do. And I want to make it very, very clear that I am not against funding in a major way the program to get to permanent storage in the United States.

I am also in favor of funding some interim storage, and doing that quickly. So I guess what I want, while I am only one Senator,

I chair the subcommittee, look, I am not even going to support the current program through another funding cycle. I mean, if it is bad, and it is, I am not going to support anything the next time because we are getting nowhere. And, Mr. Dreyfus, you have done a remarkable job with a mess, both a legislative mess and whatever other kind of mess one could describe.

But I do not believe we are going to get there on a permanent storage in the time frame that people are talking about, and therefore I think we need interim storage. And I cannot understand why there is such a delay on interim storage. Now, we have talked about that all morning, and people have beat up on everyone, and I promised myself that I did not have to do that, and I will not. I have a lot of opportunity to argue. I am going to be arguing for the next 5 days incessantly, day and night. But let me make sure that nobody misunderstands where this senator is as chairman of that subcommittee.

I am not turning my back on the waste problem. I am totally committed to getting it done. I believe we have science, technology, resources to get it done. But I do not think we are on a path to getting it done, and so with tight money I am not going to keep putting 400 million here and 300 million there to get nothing done.

I am pleased about one thing. Wherever the testimony evolved, whomever put the pen to it and finally put it in writing, that you have testified at least you did not say you would veto this bill. At least I did not read that. Maybe you have said it since then.

Dr. DREYFUS. I was afraid to ask her.

Secretary O'LEARY. You should read it again.

The CHAIRMAN. I think if you read the small print, Senator Domenici——

Senator DOMENICI. In any event, I think I am right. Look, you all need to be constructive, Madam Secretary, on this kind of bill, not destructive. We have got to fix this bill and get something done. I know there is an election coming and I know it is hard for Presidents to pick and choose in this area, but we really need some leadership from your end, and I hope we get it soon, because we need this kind of bill in order for us to fund the right kind of program, as I see it.

Thank you, Madam Secretary.

Thank you, Mr. Chairman.

Senator JOHNSTON. Mr. Chairman, may I get on the record just one question here? Dr. Dreyfus, it is fair to say at the current funding rates you could never get this repository licensed, not ever, is that correct?

Dr. DREYFUS. Well, it depends on what you mean by current funding rates.

Senator DOMENICI. $163.4 billion.

Dr. DREYFUS. In this fiscal year I am putting $215 million on Yucca. I can get back on a licensing track if you can give me an average of $300 million for Yucca Mountain and some relief on the licensing approach.

Senator JOHNSTON. I am saying at current funding.

Dr. DREYFUS. Not back to the old dates, but back to a program.

Senator JOHNSTON. Your current funding is $248.4, with 85 fenced, and you cannot do anything with that, so that leaves 163.4. You can never get there at that funding level, can you?

Dr. DREYFUS. The current funding for Yucca is $250 this year. The 85 is out of 400. We have a total program of 315. 250 going down will not do it.

Senator JOHNSTON. Ever.

Dr. DREYFUS. An average of 300, with the ability to work, will get us back to a license target date somewhere out around 5 years delay. I can shorten that up if I had more money. But that is basically it.

Secretary O'LEARY. Mr. Chairman, before Senator Domenici leaves, I understand how easy it would be to stay silent, but I would encourage his reread of the testimony that I have presented and that I have presented to this committee. I am very clear to state the President cannot support S. 1271. That, to me, implies and means that unless we can come to some negotiation that does not force us to a decision on interim storage before we can make some meaningful decision on the geologic repository, this is not legislation we will support. And I needed to be clear on that.

The CHAIRMAN. This is an opportunity for the last word.

Senator DOMENICI. You could have let that pass, and I could have been out of here.

[Laughter.]

Senator DOMENICI. I knew it anyway, but I was being generous.

The CHAIRMAN. With that generosity, we will conclude the testimony of the Secretary, and we wish you a good day, and we look forward to working with your staff as we address an expeditious process to pick up that year.

Our next witness will be Mr. Robert Pirie, Assistant Secretary for Installations and Environment for the Department of the Navy, and we will look forward to your statement relative to the ability of the Navy to continue to operate its mission faced with the reality that the Federal Government has not lived up to its commitment to the State of Idaho. I do not know whether it is the committee that is going to shut down the Navy or the State of Idaho or whether you have another plan, but we look forward to hearing your testimony and would appreciate whatever you care to submit for the record, and ask that you proceed.

STATEMENT OF ROBERT B. PIRIE, JR., ASSISTANT SECRETARY FOR INSTALLATIONS AND ENVIRONMENT, DEPARTMENT OF THE NAVY

Mr. PIRIE. Thank you, Mr. Chairman. I have submitted a fairly lengthy statement for the record, and if you would like I can summarize that briefly and spare the committee a little bit of time.

The CHAIRMAN. We appreciate that.

Mr. PIRIE. Let me just go over the major points, I think all of which or most of which have been made already. The Navy operates nuclear ships as an essential part of our force structure. Nuclear power offers unique advantages to certain classes of ships that are important to our mission. The operation of nuclear ships produces spent fuel, which has to be disposed of somehow. Since 1957, we have been sending the spent fuel for examination to the

Idaho National Engineering Laboratory, and that examination is very important. It has enabled us to improve the reliability of the nuclear fuel, and also to make technological advances which reduce the waste stream from operating our nuclear fleet.

In 1992, the Department of Energy decided to stop reprocessing the fuel that we had been shipping to Idaho, and at that point the process that Governor Batt described, in which first Governor Andrus and then he raised objections to the accumulation of the fuel in Idaho began, and that began a series of negotiations which he also described, and I think his explanation of the process was completely adequate, by which the Navy has and the Department of Energy have agreed to remove all of the fuel from Idaho by 2035. So it has to go somewhere.

I should also point out that the amount of Navy spent fuel in Idaho is an extremely small fraction of the total amount there, less than 5 percent now. When all of the fuel that is to be produced and accumulated, Navy and non-Navy, is wherever it is going to be in 2035, the Navy's amount will be a minuscule fraction of 1 percent of all that fuel. So it is not much, but it is important to the Navy, and it is important to national security.

The process that led to the need to strike an agreement with Idaho interrupted the flow of spent fuel from the refueling activities at Navy shipyards and essentially blocked up the train of available containers that moved the fuel from shipyards to Idaho, to the extent that we were faced with delaying the refueling of important ships, including the aircraft carrier U.S.S. *Nimitz.* So this is quite important to us. It is extremely important for us to honor our obligation with the State of Idaho to remove the fuel from there by 2035, and we certainly support the Department of Energy in its efforts to find the solution to that problem.

How the Department of Energy solves that problem is not our province. It is a technical problem for them to address. We just need some place to move fuel, as we have agreed to do.

[The prepared statement of Mr. Pirie follows:]

PREPARED STATEMENT OF ROBERT B. PIRIE, JR., ASSISTANT SECRETARY FOR INSTALLATIONS AND ENVIRONMENT, DEPARTMENT OF THE NAVY

INTRODUCTION

Mr. Chairman, I am the Honorable Robert B. Pirie, Jr., Assistant Secretary of the Navy (Installations and Environment). I am very pleased to have this opportunity to testify today on behalf of the Department of the Navy with regard to S. 1271, the Nuclear Waste Policy Act of 1995.

The Navy has a substantial interest in the subject of this legislation. Our interest arises because more than forty percent of the Navy's warships are nuclear powered. They include eight of the Navy's twelve current aircraft carriers, two additional nuclear powered aircraft carriers which are under construction, more than 90 nuclear powered submarines, and four guided missile cruisers.

Since the 1950s when Admiral Hyman G. Rickover brought our nuclear powered fleet into existence, nuclear powered warships have steamed safely for more than one hundred million miles and accumulated over 4,600 reactor-years of safe operation—an uninterrupted record of safety that President Clinton commended in April 1994. Our nuclear powered warships played a crucial role in winning the Cold War, and remain a vital element of the Navy's ability to fulfill its national security mission in support of our Nation's defense.

Before presenting the Navy's views on the legislation, I would like to describe the nature of the Navy's use of nuclear fuel.

NAVAL SPENT FUEL CYCLE

The Navy creates spent nuclear fuel through the operation of its nuclear powered warships and training reactors. When a warship is refueled and overhauled for continued service or is defueled because it is being inactivated, its naval spent fuel is removed at the shipyard. Similarly, naval spent fuel is removed from afloat and land-based naval training reactors when they are refueled or deactivated. In all cases, the naval spent fuel is then transported to the Idaho National Engineering Laboratory (INEL) in Idaho, which is a Department of Energy (DOE) facility.

At INEL, all of the naval spent fuel is examined at the Naval Reactors Facility (NRF) within the INEL site, and which is operated by the Naval Nuclear Propulsion Program. Examination of naval spent fuel is necessary to ensure that maximum performance and use is obtained from the naval nuclear fuel in the reactors that power the nuclear powered warships which are currently serving with the fleet. Naval spent fuel examination is also necessary to enable naval fuel having longer lifetimes to be designed. The first nuclear powered warship, USS NAUTILUS, required refueling after two years of operation. Today, as a result of the technical advances in which examination of naval spent fuel has played a pivotal role, nuclear powered warships operate more than twenty years without refueling. For the next generation of attack submarines, the Navy is developing nuclear fuel which will last as long as the life of the ship, about 30 years. This progress means that far less naval spent fuel will be created in propelling our nuclear powered fleet in the future.

The facilities at NRF for naval spent fuel examination are unique. Examinations have been conducted safely there since 1957. There is no other facility in the United States which can perform this critical function. It would take approximately eight years and cost the taxpayers approximately $800 million to design and construct an alternate facility which would provide the same level of performance, safety, and environmental protection at a site having different seismic and environmental conditions.

Before 1992, naval spent fuel that had been examined was reprocessed by DOE to recover uranium for other uses. In 1992, DOE ceased reprocessing all spent fuel, including naval spent fuel. Since then, naval spent fuel has been stored at INEL after its examination pending final disposition.

Because naval reactors are relatively small and no longer require frequent refuelings, the total amount of naval spent fuel generated is very small. Using the unit of measure prescribed in the Nuclear Waste Policy Act of 1987 and in Nuclear Regulatory Commission regulations, there are about 12 metric tons of heavy metal (which I will refer to as metric tons) of naval spent fuel at INEL today. By comparison, about 250 metric tons of non-naval spent fuel are also stored at INEL. Thus, naval spent fuel constitutes less than five percent of the spent fuel at INEL. For further comparison, about 30,000 metric tons of commercial spent fuel are in storage throughout the U.S. today. Naval spent fuel constitutes less than one twentieth of one percent of that amount. Counting the 12 metric tons already at INEL, the Navy will have a total of 65 metric tons of naval spent fuel by the year 2035. By that same date, there will be more than 80,000 metric tons of commercial spent fuel and more than 2,700 metric tons of DOE spent fuel throughout the nation. Thus, naval spent fuel will amount to less than one tenth of one percent of the total amount of spent fuel which the proposed legislation addresses. As I have noted above, however, the nuclear powered warships whose use creates naval spent fuel are essential components of the Navy's ability to meet its national security obligations.

NATURE OF NAVAL NUCLEAR FUEL

Naval nuclear fuel is designed to withstand exceptionally challenging and adverse combat conditions. It must also operate safely in close proximity to the sailors who live, eat and work every day, and often in the case of deployed submarines for months at a time, within the confines of their warship. As an illustration of its ruggedness and high integrity, naval nuclear fuel can:

- Withstand, without damage, combat shock loads well in excess of 50 times the force of gravity, which is more than 100 times the force experienced in a severe earthquake;
- Fully contain within the fuel all "fission products"—the radioactive materials created in the fuel as a consequence of the operation of a nuclear reactor;
- Sustain, without damage, repeated power transients from low to high power or vice versa, each occurring within a short time, as required to allow the warship to adjust its speed;
- Operate for more than 20 years without refueling. For example, USS NIMITZ went to sea in 1975 and will receive her first and only refueling in 1998.

These characteristics of naval fuel, which are necessary to ensure that nuclear powered warships can fulfill their military mission, also make the fuel exceptionally well-suited for safe transport, storage and ultimate disposal after service.

INEL AND THE IDAHO AGREEMENT

The Navy has shipped its naval spent fuel to NRF at INEL for examination since 1957. All of the 631 container shipments have been made safely. The shipments are performed by rail using specially built and certified shipping containers which meet stringent Navy, DOE, Department of Transportation, and Nuclear Regulatory Commission requirements. One example of the rugged design for our most recent container, the M-140, is that the naval spent fuel it carries is enclosed within a cylinder having a wall thickness of 14 inches of stainless steel, which is greater than that found in a battleship turret.

In June 1993, the Federal District Court for the District of Idaho enjoined shipments of naval and DOE spent fuel to INEL until a comprehensive Environmental Impact Statement was prepared covering future management of that spent fuel. The injunction was amended in December 1993 to allow certain shipments of naval spent fuel to avoid restricting the Navy's ability to refuel and defuel nuclear powered warships.

The spent fuel EIS, which includes 5,200 pages of evaluation and analysis and cost more than $50 million dollars to prepare, covers the next 40 years of Navy and DOE spent fuel management. The EIS concluded that naval spent fuel could be safely transported to INEL, examined at NRF and then stored and managed at INEL during that period without significant impact on health or the environment. The EIS was published in April 1995 and a Record of Decision issued in June 1995.

Following issuance of the Record of Decision, the District Court elected to continue the injunction while considering the adequacy of the EIS, in view of concerns expressed by the State of Idaho. During the four month period between June and October 1995, the Navy, DOE, and the Governor and State of Idaho worked to achieve a negotiated settlement. The federal and State parties agreed that the activities to be performed by them would be in the public interest. Accordingly, their settlement was submitted to the Court for its consideration. On October 17, 1995, the Court so ordered the agreement in its entirety.

Among its provisions, the Court's order assures that the Navy will be able to ship its spent nuclear fuel to INEL for the next forty years to support refueling and defueling of naval warships and examination of naval spent fuel. This is an essential predicate for the uninterrupted operation of the Navy's nuclear powered fleet. The settlement also provides for extensive upgrades, management and environmental remediation work concerning naval spent fuel at INEL. For example, naval spent fuel will be moved from wet storage in water pools to dry storage in aboveground containers. A further provision requires naval spent fuel to be removed from INEL to a permanent geologic repository or interim storage site outside Idaho by the year 2035, and, subject to the availability of appropriations provided in advance for that purpose, directs the Navy to make substantial payments to Idaho for every day that deadline is not met.

The Court's order removes a significant threat to national security. In connection with the litigation, the Secretary of Defense certified that national security was threatened by the Navy's inability to transport its naval spent fuel to INEL. In that regard, the Chief of Naval Operations and the Director of the Naval Nuclear Propulsion Program testified that the Navy was suffering irrevocable harm every day that the interruption in its shipments by Idaho prevented the examination of its spent fuel and the refueling or defueling of its warships. The Court's order has enabled the Navy to resume its schedule for operating, refueling and defueling our nuclear powered fleet and examining naval spent fuel without further delay from Idaho or its lawsuit.

In order to continue to receive the benefits of the Court order, the Navy must fully perform its obligations. Paramount in that respect is the requirement that naval spent fuel be removed from Idaho by the year 2035. To accomplish this, a permanent geologic repository or interim storage facility must be operational outside Idaho by that date to receive the naval spent fuel. Moreover, in the words of the Court's order, the naval spent fuel that is stored at INEL on the date of the opening of a permanent repository or interim storage facility must be among the early shipments of spent fuel to go there.

In this regard, it is important to be clear about the Navy's interest and position. To enable the Navy to comply with the Court's order, a permanent repository or interim storage facility should be brought into existence outside Idaho as soon as practicable, and in no event later than a date which will permit the naval that spent fuel to be removed from Idaho by 2035. Beyond that goal, however, the Navy is not in a position to select or recommend a location for that facility. Nor would we recommend how the location should be selected, the criteria for the selection, or what schedule would be appropriate for its licensing and construction.

Within those parameters, the Navy supports efforts to site, license and construct a permanent repository or interim storage facility using sound principles of engineering, safety, and environmental protection in accordance with established legal procedures, including the opportunity for public involvement. We are confident that naval spent fuel is technically suitable for storage or disposal, and we will work to be ready to move naval spent fuel from to the permanent repository or interim storage facility once it is available to accept naval spent fuel.

CONCLUSION

The Navy is committed to fulfilling its obligations under the settlement agreement with the State of Idaho and the Court's order. The proposed legislation contains elements which would assist the Navy in doing so. It also contains elements which we understand are of concern to our colleagues at DOE, the Department of Justice, and other federal agencies. Our testimony today is not intended to speak to concerns that other agencies may have about this bill or to express the views of the Administration on the bill as a whole. We appreciate the opportunity to explain the Navy's need for a permanent repository or interim storage by the year 2035, which will enable the Navy to meet its obligations within the framework of responsible environmental planning, our national security, and the equities of each affected state.

The CHAIRMAN. Senator Johnston.

Senator JOHNSTON. Mr. Chairman, I have read Secretary Pirie's statement in full. It is an excellent statement. I have no questions.

The CHAIRMAN. Senator Craig.

Senator CRAIG. Mr. Chairman, thank you very much.

Mr. Secretary, on page 5 of your testimony you make some—first of all, I should not say it in that context. I appreciate your whole testimony. On page 5 you make several statements that I think are extremely important. One of them says in order to continue to receive the benefits of the court order, the Navy must fully perform its obligations. If it failed to perform its obligations, what could the State of Idaho do under the court order?

Mr. PIRIE. I believe that the provisions of the court order are for a fine of $60,000 a day for every day after the date in 2035 that our fuel was not removed from Idaho.

Senator CRAIG. But up until that point, the State can actually again—or the court could actually again stop the movement of Navy spent fuel, could it not, for failure to perform?

Mr. PIRIE. I am certain that if we do not make substantial progress between now and 2035 that Idaho will raise the issue again, and I assume that will find its way into the courts.

Senator CRAIG. But 2035 is critical, and you go on to talk about a permanent repository or interim storage facilities should be brought into existence outside Idaho as soon as practicable, and in no event later than a date which will permit the naval shipment to be removed from Idaho by 2035. You have been patient all morning. You have listened to the Secretary of Energy. You have said it is not our job to find the spot, we have just simply committed to move it.

Mr. PIRIE. Yes, sir.

Senator CRAIG. Do you not think the Secretary of Defense ought to talk to this Secretary, after what you have just heard, and tell her how important it is, or the nuclear Navy could get dry-docked?

Mr. PIRIE. I think we have ample opportunities to underline our concern to the Department of Energy, yes, sir.

Senator CRAIG. Because I see a little bit of schizophrenia going on here, and this is not to point fingers at you because you have operated in good faith with the State of Idaho, and I know the Governor appreciates that, and the negotiations that went forward were well handled by the Navy, and I respect that. But I think it is important that the Secretary of Defense—therefore, the Secretary of the Navy—lift this issue to a higher priority than apparently this administration is willing to give it. That is my point.

I would only suggest to you that that needs to get done, and if it were a team effort there is a strong likelihood that it would get done because the agreement that was struck with Idaho was a team effort: Idaho, Navy, and DOE. But now it has been lifted to the top of the stack, and I guess that would only be my observation, at least from your testimony, Mr. Secretary, that that would be critical.

Mr. PIRIE. Thank you, Senator.

Senator CRAIG. Thank you.

The CHAIRMAN. Thank you very much. I think we have no further questions, and would wish you a good day. Your participation, while limited here, was certainly significant, and you got a drift of where this thing is relative to the other witnesses.

Mr. PIRIE. I am always glad to be any help I can to the committee.

The CHAIRMAN. Thank you very much.

We are going to conclude with the next panel, which consists of Gary Gates, vice president for Nuclear Operations for the Omaha Public Power District, on behalf of the American Public Power Association; followed by Mr. Bill Sherman, Vermont Department of Public Services and cochair of the Northeast High-Level Radioactive Waste Transportation Task Force. That is quite a mouthful, and we understand, Mr. Sherman, that Vermont has extraordinary deposits of granite which make probably the best storage from the standpoint of the type of material.

If you ideally went to look for a place for a permanent or a temporary repository, why, the granite that you have in your State—and this was outlined in great detail by the Swedes who kept asking us "why do we not go to Vermont?" Because that is where you have the best——

Mr. SHERMAN. Yes, Mr. Chairman, all very fractured granite, though. Very fractured.

[Laughter.]

The CHAIRMAN. In any event, I wanted to at least note that Vermont does come up occasionally as a potential site because of your granite. And Mr. Dick Snell, chairman and CEO of Pinnacle West Capital Corporation. We welcome you to the committee, and we understand that you have a capability to burn MO_x fuel, given the opportunity and the authority and so on. This is rather interesting, as we determine ultimately what to do with this waste and whether to put it in a permanent repository or move into the technology that is clearly developing as the alternative, as evidenced by the French and the Japanese. The French have a mandate that they simply do not or cannot bury or put away plutonium. They reprocess it and burn it, and as a consequence reduce the proliferation threat, which makes an awful lot of sense to me as we discuss the ultimate disposition of this material.

And Mr. Corbin McNeill, president and CEO of PECO Energy Company. We look forward to your testimony.

Gentlemen, it is 10 after lunchtime, so anybody that wants to abbreviate their statement, it would be appreciated. So please proceed.

Gentlemen, you have heard the testimony, so you know where the committee is and you know where the witnesses are. The press has pretty much left us so, you know, they are, again, not going to get the full story, but that does not surprise anybody. That is with the exception of those that are still here.

Please proceed.

STATEMENT OF GARY GATES, VICE PRESIDENT FOR NUCLEAR OPERATIONS, OMAHA PUBLIC POWER DISTRICT, ON BEHALF OF THE AMERICAN PUBLIC POWER ASSOCIATION

Mr. GATES. Thank you, Mr. Chairman. I will abbreviate my remarks in the interest of time.

Mr. Chairman and members of the committee, thank you for the opportunity to appear today on behalf of the American Public Power Association, APPA. As indicated, my name is Gary Gates. I am vice president of Nuclear Operations for Omaha Public Power District. OPPD is a political subdivision of the State of Nebraska, governed by a publicly elected board of directors. While our headquarters are located in Omaha, Nebraska, our service territory encompasses the eastern-most 13 counties of the State of Nebraska, including the greater metropolitan area of the city of Omaha.

OPPD serves the electrical needs of more than 600,000 people, 270,000 customer-owners, through a balanced mix of fossil and nuclear generation. Approximately 40 percent of this energy is supplied by the Fort Calhoun Nuclear Station.

APPA is the national service organization representing the interests of over 2000 public, municipal, and other State and local government-owned utilities throughout the United States. In addition to the city of Omaha, APPA member utilities serve some of the Nation's largest cities such as Los Angeles, Sacramento, Seattle, Phoenix, Jacksonville, Austin, San Antonio, Nashville, Memphis, Cleveland, and Orlando. Several State public power agencies, such as New York Power Authority, South Carolina Public Service Authority, and Lower Colorado River Authority in Texas, provide electric power to many communities within their States.

However, the majority of APPA members are located in small- and medium-sized communities in every State except Hawaii. Collectively, public power utilities deliver electric energy to one of every seven U.S. electric customers, or about 35 million people. Public power represents the efforts of local governments to shape and improve their economic future through the provision of affordable electricity.

I am pleased to share APPA's support for the enactment of S. 1271 to provide a central interim storage facility for high-level nuclear waste. The Nuclear Waste Policy Act of 1982 required the Department of Energy to construct and operate a permanent storage facility for high-level nuclear waste by 1998. DOE estimated in

1989 that a permanent storage facility will not be in operation until at least 2010. Many of the Nation's 109 nuclear powerplants, including my company's Fort Calhoun Nuclear Station, will run out of onsite storage capacity long before this target date. In light of this significant delay, Congress must act to authorize an interim storage facility and to ensure this vitally important program is brought back on track.

If enacted, S. 1271 would allow for operation of a storage facility by the target date established in the Nuclear Waste Policy Act of 1982, and would lay the groundwork for future operation of a permanent site. Most importantly, the bill accomplishes these objectives by providing an integrated approach to nuclear waste management through a comprehensive transportation plan and requirements for spent fuel container technology.

Since 1982, electric consumers have made significant financial contributions toward a responsible solution. The nuclear waste fund was created by the Act to collect fees from consumers of nuclear energy to cover the cost of constructing and operating permanent storage facilities. To date over $10 billion has been collected to address this issue. In my home State of Nebraska, consumers have paid a staggering $136 million in the nuclear waste fund, yet while consumers have met their obligation the Federal Government has failed to hold up its end of the deal. As a result, utilities and their consumers must either construct new onsite storage facilities or face possible power shortages and drastically increased electric rates if plants are required to shut down due to inadequate storage for spent fuel.

For example——

The CHAIRMAN. I wonder if you could summarize. We have your statement.

Mr. GATES. Just to close, then, with the impact on my company, to date we have contributed over $58 million to the nuclear waste fund, which is approximately $3 million annually. To the end of its life we will contribute $110 million. With an interim storage facility available in 1998, we could operate Fort Calhoun to be under this to the end of its life with an additional cost of $117 million. Keep in mind, the original cost of our station was only $178 million.

Consumers and local governments have carried their share of responsibility for furthering sound public policy, and enacting S. 1271 is a necessary step toward getting the Federal Government's waste program back on track.

Thank you.

[The prepared statement of Mr. Gates follows:]

PREPARED STATEMENT OF GARY GATES, VICE PRESIDENT FOR NUCLEAR OPERATIONS, OMAHA PUBLIC POWER DISTRICT

Mr. Chairman, and members of the committee, thank you for the opportunity to appear today on behalf of the American Public Power Association (APPA). My name is Gary Gates, and I am the Vice President for Nuclear Operations for the Omaha Public Power District (OPPD).

OPPD is a political subdivision of the State of Nebraska, governed by a publicly-elected Board of Directors. While our headquarters are located in Omaha, Nebraska, our service territory encompasses the eastern-most 13 counties of the State of Nebraska, including the greater metropolitan area of the City of Omaha. OPPD serves the electrical needs of more than 600,000 people (270,000 customer-owners) through a balanced mix of fossil and nuclear generation. Approximately 40% of this energy is supplied by our Fort Calhoun Nuclear Station.

APPA is the national service organization representing the interests of over 2,000 public, municipal, and other state and local government-owned utilities throughout the United States. In addition to the City of Omaha, APPA member utilities serve some of the nation's largest cities, such as Los Angeles, Sacramento, Seattle, Phoenix (Salt River Project), Jacksonville, Austin, San Antonio, Nashville, Memphis, Cleveland, and Orlando. Several state public power agencies, such as New York Power Authority, South Carolina Public Service Authority, and Lower Colorado River Authority in Texas provide electric power to many communities within their states. However, the majority of APPA members are located in small and medium-sized communities in every state except Hawaii. Collectively, public power utilities deliver electric energy to one of every seven U.S. electric consumers, or about 35 million people. Public power represents the efforts of local governments to shape and improve their economic future through the provision of affordable electricity.

I am pleased to share APPA's support for enactment of S. 1271 to provide a central interim storage facility for high-level nuclear waste. As previously mentioned, Omaha Public Power District owns and operates the Fort Calhoun Nuclear Station located approximately 20 miles north of Omaha near Blair, Nebraska. Because we project the plant's on-site storage capacity for spent nuclear fuel will be exhausted by the year 2006, we have a very strong interest in the operation of an interim storage facility by 1998. In the absence of a safe and centrally-located interim facility, OPPD will at best be faced with the prospect of providing additional on-site storage measures or, at worst, having to discontinue operation of the plant prior to its licensed operational life.

More broadly, failure to enact legislation that would allow construction and operation of an interim storage facility by 1998 threatens our nation's nuclear generation capacity and unfairly requires consumers and local governments to bear the cost burden for bureaucratic delays in this program area. APPA strongly supports S. 1271, as it would enable the federal government to fulfill its commitment to ratepayers, while providing for an integrated and responsible approach to high-level nuclear waste management.

A COMPREHENSIVE APPROACH IS NEEDED

The Nuclear Waste Policy Act of 1982 required the Department of Energy (DOE) to construct and operate a permanent storage facility for high-level nuclear waste by 1998. DOE estimated in 1989 that a permanent storage facility will not be in operation until at least 2010. Many of the nation's 109 nuclear plants, including OPPD's Fort Calhoun Station, will run out of on-site storage capacity long before this target date. In light of this significant delay, Congress must act to authorize an interim storage facility and ensure this vitally important program is brought back on track. If enacted this Congress, S. 1271 would allow for operation of a storage facility by the target date established in the Nuclear Waste Policy Act of 1982, and would lay the groundwork for future operation of a permanent site. Most importantly, S. 1271 accomplishes these objectives by providing an integrated approach to nuclear waste management through a comprehensive transportation plan and requirements for spent fuel container technology.

In an effort to identify a solution to this issue, several municipally-operated utilities and the all public power state of Nebraska have intervened in a case (*State of Michigan, et al.* v. *DOE*) pending before the U.S. Court of Appeals for the District of Columbia asking that DOE be obligated to meet its 1998 timeline for acceptance of spent nuclear fuel. However, given the complexity and importance of the high-level nuclear waste storage issue, APPA believes it is more appropriate for Congress, not the courts, to approach the problems raised by the delays in the program. Much-needed deliberation and review of the public policy considerations related to high-level waste storage will not be provided if the issue is simply left to the courts to decide.

CONSUMERS HAVE A STAKE IN S. 1271

Since 1982, electric consumers have made significant financial contributions toward a responsible solution to the need for a high-level waste storage facility. The Nuclear Waste Fund was created by the Nuclear Waste Policy Act of 1982 to collect fees from consumers of nuclear energy to cover the costs of constructing and operating a permanent storage facility for nuclear waste. To date, over $10 billion has been collected to enable DOE to effectively address this important issue. In my home state of Nebraska, consumers have paid a staggering $136 million into the Nuclear Waste Fund.

Yet, while consumers have met their obligation, the federal government has failed to hold up its end of the deal. As a result, utilities and their customers must either construct new on-site storage facilities, or face possible power shortages and drastically increased electricity rates if plants are required to shut down due to inadequate storage for spent fuel. For example, at our Fort Calhoun Station, OPPD has already spent nearly $7 million to provide for additional on-site storage. In effect, the consumers of nuclear energy are being double-taxed. A significant amount of the funding they have provided to support DOE's high-level waste storage program has been redirected toward general deficit reduction—and they are left to find alternative solutions to the problem at a very high cost.

Sharing these concerns regarding management of the Nuclear Waste Fund, the Consumer Federation of America has also taken a strong position urging Congress and the Administration to "ensure that the billions of dollars already collected from consumers, as well as the funds that will be collected in the future, are used for purposes specified in the Nuclear Waste Policy Act and not retained in the Treasury for deficit reduction."

PUBLIC SAFETY GOALS ARE FURTHERED BY S. 1271

In comparison with on-site storage of spent fuel at nuclear facilities, a central storage location allows for an increased margin of public safety and operating efficiency. Nuclear Regulatory Commission and Department of Energy regulators and inspectors can more efficiently conduct surveillance and inspections of one centralized site. Moreover, electric consumers will be provided relief as a central security plan can be implemented instead of costly and duplicative security measures taken at plants across the country—including active security forces at decommissioned plants due to the presence of high-level nuclear waste at these facilities. As safe as our facilities are today, a centralized facility will be even safer.

The remote location of an interim, and very likely a permanent storage facility, in Yucca Mountain, Nevada provides additional security and environmental advantages as spent fuel can be more effectively managed at a single, secure site with state-of-the art technology. Recognizing these benefits, Congress authorized site characterization for a permanent facility at Yucca Mountain in 1987.

In addition, advancement in transportation technology for nuclear fuel has ensured that nuclear fuel can be safely shipped to a central facility. First, the design of containers to be used for transport has been carefully regulated by the Nuclear Regulatory Commission, and extensive testing has indicated the design meets the highest possible standards to ensure public safety. In addition, spent fuel represents only a small portion of the overall hazardous material shipments in the U.S. on an annual basis. Routes used for transport are specifically designated by the U.S. Department of Transportation, and DOE plays an important role in training state and local officials in the development of emergency response plans. Finally, the transportation of nuclear waste has a proven record of safety. The public has never been placed at risk due to a release of radioactivity from the transportation of high-level nuclear waste.

THE OPPD EXPERIENCE

Before closing my testimony on behalf of APPA, let me illustrate the significance of this issue to OPPD:

- Our Fort Calhoun Station, a 476 MW PWR, went into commercial operation in September of 1973, at an original cost of $178 million. I would ask that you please keep that figure in the back of your mind for a few moments.
- To date, OPPD has contributed over $58 million to the Nuclear Waste Fund—approximately $3 million annually.
- At the end of the licensed life of the Fort Calhoun Station in 2013, our contribution will have grown to nearly $110 million.
- In order to continue to provide for on-site storage of spent nuclear fuel (absent a Federal repository), we have been forced into "reracking" of our spent fuel pool at a cost of nearly $7 million.
- The most recent rerack in 1994 provides OPPD on-site storage capacity only until approximately 2006—absent new technology, no further reracks are feasible.
- Absent an interim storage facility, OPPD would then be forced into on-site dry cask storage which, in addition to high costs, has also been met with considerable political opposition.
- With an interim storage facility available in 1998 (and per the acceptance schedule for receipt of spent nuclear fuel), OPPD could operate the Fort Cal-

houn Station to its licensed life without additional costs above the $117,000,000 already mentioned.

- In light of these figures, again consider that our cost to construct the Fort Calhoun Station was $178,000,000. It is mind-boggling that the amount of money spent by our customer-owners to address spent fuel storage is quickly approaching the amount of our original investment in the plant.

PUBLICLY-OWNED UTILITIES SUPPORT CONGRESSIONAL ACTION ON S. 1271

America's municipally-owned and operated utilities urge Congress to act expeditiously in its consideration of S. 1271. Consumers and local governments have carried their share of responsibility for furthering sound public policy in the area of nuclear waste management. Enacting S. 1271 is a necessary step toward getting the federal government's nuclear waste storage program back on track and working in the public interest.

APPA MEMBER UTILITIES CONTRIBUTING TO THE NUCLEAR WASTE FUND

[Prepared by the American Public Power Association, December 1995]

Publicly owned utility by State	Plant	Status	Share [percent]	Year plant will run out of waste storage space	Paid into Department of Energy Nuclear Waste Fund
Salt River Project (AZ)	Palo Verde 1	CO (1986)	17.49	2005	$27,400,000
	Palo Verde 2	CO (1986)	17.49	2005	
	Palo Verde 3	CO (1988)	17.49	2005	
City of Anaheim Public Utilities (CA)	San Onofre 2	CO (1983)	3.16	2001	4,300,000
	San Onofre 3	CO (1984)	3.16	2003	
Los Angeles Department of Water and Power (CA)	Palo Verde 1	CO (1986)	5.7	2005	22,000,000
	Palo Verde 2	CO (1986)	5.7	2005	
	Palo Verde 3	CO (1988)	5.7	2005	
City of Riverside Public Utilities (CA)	San Onofre 2	CO (1983)	1.79	2003	3,000,000
	San Onofre 3	CO (1984)	1.79	2004	
Sacramento Municipal Utilities District (CA)	Rancho Seco 1 closed in 1989	CO (1975)	100	[1]	40,200,000
Southern California Public Power Authority (CA)	Palo Verde 1	CO (1986)	5.91	2005	30,500,000
	Palo Verde 2	CO (1986)	5.91	2005	
	Palo Verde 3	CO (1988)	5.91	2005	
Florida Municipal Power Agency (FL)	St. Lucie 2	CO (1983)	8.81	2001	5,317,680
Orlando Utilities Commission (FL)	St. Lucie 2	CO (1983)	6.08	2001	1,200,000
	Crystal River 3	CO (1977)	1.6015	2010	
Dalton Utilities (GA)	Vogtle 1	CO (1987)	1.6	2009	
	Vogtle 2	CO (1989)	1.6	2009	
	Hatch 1	CO (1975)	2.2	2003	
	Hatch 2	CO (1979)	2.2	2003	
Municipal Electric Authority of Georgia (GA)	Vogtle 1	CO (1987)	22.7	2209	43,248,255
	Vogtle 2	CO (1989)	22.7	2009	
	Hatch 1	CO (1975)	17.7	2003	
	Hatch 2	CO (1979)	17.7	2003	
Massachusetts Municipal Wholesale Electric Company (MA)	Seabrook 1	CO (1990)	11.59	[5] 2010	305,450
	Millstone 3	CO (1986)	4.8	[5] 2010	
Taunton Municipal Lighting Plant (MA)	Seabrook 1	CO (1990)	.10	[5] 2010	1,534,702
	Maine Yankee	CO (1972)	.53	2008	
	Vermont Yankee	CO (1972)	.46	2005	

APPA MEMBER UTILITIES CONTRIBUTING TO THE NUCLEAR WASTE FUND—Continued

[Prepared by the American Public Power Association, December 1995]

Publicly owned utility by State	Plant	Status	Share [percent]	Year plant will run out of waste storage space	Paid into Department of Energy Nuclear Waste Fund
Nebraska Public Power District (NE)	Cooper	CO (1974)	100	2004	82,500,000
Omaha Public Power District (NE)	Fort Calhoun 1	CO (1973)	100	2006	56,000,000
New York Power Authority (NY)	Indian Point 3	CO (1976)	100	2008	93,900,000
	Fitzpatrick	CO (1975)	100	2002	
North Carolina Eastern Municipal Power Agency (NC)	Brunswick 1	CO (1977)	18.33	[2] 2000	2,000,000
	Brunswick 2	CO (1975)	18.33	[2] 2000	
	Shearon Harris 1	CO (1987)	16.17	[3]	
North Carolina Municipal Power Agency #1	Catawba 2	CO (1986)	75	[4] 2006	40,500,000
Eugene Water and Electric Board (OR)	Trojan retired 1/92	CO (1976)	30	[1]	34,500,000
Piedmont Municipal Power (SC)	Catawba 2	CO (1986)	25	[4] 2006	15,625,000
South Carolina Public Service (SC)	V.C. Summer 1	CO (1984)	33.33	2008	21,000,000
Austin Electric Utilities (TX)	South Texas 1	CO (1988)	16	2020	12,051,387
	South Texas 2	CO (1989)	16	2021	
San Antonio Public Service Board (TX)	South Texas 1	CO (1988)	28	2020	
	South Texas 2	CO (1989)	28	2021	
Washington Public Power Supply System (WA)	WA Nuclear Plant 2	CO (1984)	100	2000	57,222,494
Federal Public Power Systems:					
Tennessee Valley Authority	Browns Ferry 1 closed in 1985	CO (1974)	100	2007	297,400,000
	Browns Ferry 2	CO (1975)	100	2007	
	Browns Ferry 3 closed in 1985 reopening in 1996	CO (1977)	100	2007	
	Sequoyah 1	CO (1981)	100	2003	
	Sequoyah 2	CO (1982)	100	2003	
	Watts Bar 1	CP (1996)	100	2020	

[1] Plant closed, date cannot be projected. [2] After this date, waste will be stored at the Harris plant. [3] Date cannot be projected. [4] With re-racking, storage space could be available until beyond 2020. [5] After re-racking.

CO = commercial operation. CP = construction permit.

The CHAIRMAN. Thank you, Mr. Gates.
Mr. Sherman.

STATEMENT OF WILLIAM K. SHERMAN, ON BEHALF OF THE VERMONT DEPARTMENT OF PUBLIC SERVICE, AND THE NORTHEAST HIGH-LEVEL RADIOACTIVE WASTE TRANSPORTATION TASK FORCE

Mr. SHERMAN. Thank you, Mr. Chairman. members of the committee, ladies and gentlemen, I am Bill Sherman, and I am from the Vermont Department of Public Service. And in that regard I bring greetings from our Governor Dean, who sends his regret that a prior engagement kept him from being here today on the panel. I also represent the Northeast high-level radioactive waste transportation task force, of which I am cochair. Thank you for the opportunity for appearing and making a statement. I ask that my written comments be incorporated into the record appropriately.

On behalf of the transportation task force, we are a task force of government appointees organized by the Council of State Governments Eastern Regional Conference. We represent 10 Northeastern States from Delaware and New Jersey, Pennsylvania, up through the Northeast. At our meeting this fall I think we passed a resolution which should be of interest to the committee. Our resolution was a strong statement passed unanimously. Here is what it said. It had two points.

Transportation is a very important issue which needs careful planning, training, and implementation. That is the first item. The second item is this, that transportation with careful planning, training, and implementation, can be done safely and should not be considered an impediment to centralized interim storage as considered by your bill, Senator Craig.

Now, who are we that make this statement? Senators, we are the experts in the Northeast. We are the ones who have experience with radioactive waste handling, and that is why our Governors appointed us to this task force. How can we make this statement? First, if the bill is passed as contemplated, it will enable transportation of spent fuel of an order of about four times what has occurred over the last 15 years. This is significant, but it is not an increase that is different from what we have been experiencing in the Northeast.

Spent fuel moves in the Northeast, and I have a number of examples in my written testimony. I will not cover them all, but I will mention one or two. Windsor Locks, Connecticut, just this month shipped fuel by rail through Hartford, Connecticut, Springfield, Massachusetts, and then west to Idaho Falls. Several months ago the University of Rhode Island shipped spent fuel from its reactor through Connecticut, Pennsylvania, to Savannah River, South Carolina. Moving fuel is not new in the Northeast.

Now, we are also the experts in hazardous waste movement, and we have this important statement to say, that we know that spent fuel transportation is less risky than much of the hazardous waste transportation that goes on right now. So our statement is strong. Transportation of spent fuel should not be considered an impediment for establishing centralized interim storage.

Senator JOHNSTON. Mr. Chairman, may I interrupt at this point to ask has there ever been an accident in which anyone was radiated in the United States of America in the history of this program, the transportation?

Mr. SHERMAN. No, Senator, there has not.

Senator JOHNSTON. I do not know of any worldwide, do you?

Mr. SHERMAN. I cannot answer worldwide, but I know nationally there has not.

Senator JOHNSTON. I think the answer is that there has never been. We ought to figure out how many miles that is and how many shipments. I do not think there has ever been one single one. And there is not likely ever to be.

Mr. SHERMAN. Now shifting to my role with the Vermont Department of Public Service, let me state that I believe that my comments are representative of most of the States, the majority of the States, who have electric consumers who pay for nuclear energy. Committee members, we are frustrated in the States. We are extremely frustrated by money, lots of money, paid in, hardly any result, money diverted from its intended purpose for other reasons. We feel that it needs to stop and interim storage needs to be implemented. And in my testimony I characterize it in terms of three responsible actions which I will just mention.

The first is money responsibility. The responsible choice for storage is interim storage. It costs about half-a-billion dollars every year we do not implement interim storage. And let me say this about one characterization I have heard this morning. It is easy to characterize this as a benefit to my colleagues on the panel, members from the nuclear industry, but it is the people that I represent who are paying for this, electric consumers. And electric consumers have the same faces as the people you represent, the taxpayers. It is these people who are paying an extra half-a-billion a year every time we do not implement that. So to be responsible money-wise you need to implement interim storage.

Now, let me mention environmental responsibility. Interim storage is the green solution for nuclear fuel. Let me just tell you a quick story. In Vermont, if you do not implement interim storage, there will be a time in the near future where our Vermont nuclear plant puts up dry cask storage.

Now, you all must be aware that there is a low-level waste issue that looms around as well as this high-level waste. In Vermont, we studied our site at Vermont Yankee, and we found that site inappropriate for low-level waste because anything that goes onto the ground at Vermont Yankee will quickly find its way into the Connecticut River and go down past Hartford and into Long Island Sound.

Now, we may be in a situation, Senators, where the site which was unacceptable for low-level waste, which is only really hazardous for 100 years and maybe 300 years, may be the long-term place where high-level waste goes because of the way the laws are written. There is something wrong-headed about that. The green solution is a warm, dry place that has been well studied.

Now, the last thing, and I am sorry for going long, the last responsibility is governmental responsibility. I include within my written testimony an excerpt from the Atomic Energy Act of 1946, when the Government began and the Congress began promoting nuclear power and nuclear energy. I also include a statement from my department's annual report in 1958, and let me read it, at the risk of more time.

And I quote:

One of the chief difficulties encountered, and this is about building the Yankee Row Nuclear Plant, which has now shut down and trying to decommission. One of the chief difficulties encountered seems to be a safe manner in which to dispose of waste materials that contain potentially dangerous radioactive matter. No doubt this will eventually be accomplished, but present problems in this respect appear to be considerable.

Our department said this is a problem in '58, but we went ahead with it anyway, on the assurance by the promoters, the Government, that this problem will be solved. Now here we are some 37 years later, the plant is shut down, and our ratepayers are paying without receiving any benefit for the storage of that fuel at Yankee Row. You need to act on this bill for this reason, for credibility and responsibility in Government.

We are very frustrated and desperate in the States, and I urge you very strongly to consider and enact this legislation. Thank you for the opportunity to make this statement.

[The prepared statement of Mr. Sherman follows:]

PREPARED STATEMENT OF WILLIAM K. SHERMAN, ON BEHALF OF THE VERMONT DEPARTMENT OF PUBLIC SERVICE, AND THE NORTHEAST HIGH-LEVEL RADIOACTIVE WASTE TRANSPORTATION TASK FORCE

Mr. Chairman, members of the Committee, ladies and gentlemen, I am Bill Sherman, representing the Vermont Department of Public Service. In that regard, I bring greetings from our Governor Howard Dean who sends his regrets that a prior engagement kept him from appearing here today. I am also representing the Northeast High-Level Radioactive Waste Transportation Task Force, of which I am Co-Chair. I would like to thank you for the opportunity to address the Committee.

NORTHEAST HIGH-LEVEL RADIOACTIVE WASTE TRANSPORTATION TASK FORCE

First on behalf of the Transportation Task Force: The Northeast High-Level Radioactive Waste Transportation Task Force is a shareholder group convened by the Council of State Governments—Eastern Regional Conference for the purpose of interacting the Department of Energy (DOE) on spent nuclear fuel transportation issues. The Task Force members include governor appointees in energy policy, radiological health and emergency management positions from the nine northeastern states of Connecticut, Delaware, Maine, Massachusetts, New Hampshire, New Jersey, New York, Pennsylvania and Vermont. There are 28 commercial nuclear reactors located in these states at 18 sites.

At our Fall meeting this year, we unanimously passed a strong statement about spent fuel transportation in the form of a Resolution (Attachment A)* which I believe will prove useful for the Committee's consideration. The statement has two parts. First, spent fuel transportation is a real issue requiring careful planning, training and implementation. And second, with careful planning, training and implementation, spent fuel transportation can be done safely and should not be an impediment for interim storage.

Let's consider who we are that made this statement. We are the designated officials in our representative states who are the experts for radioactive and hazardous material transportation. We have experience doing this work which is why we were appointed by our respective governors to this task force. As state administration officials, we continuously take actions which must balance the public interests within our states—environmental interests, business interests, groups opposed to nuclear power, nuclear utilities and the public at-large. None of us would be comfortable with a pro- or anti-nuclear label.

---

*All attachments have been retained in committee files.

How can we make this statement? As we state in the Resolution, a decision to establish a single interim storage facility would increase spent fuel transportation in the next 40 years by four times the rate over the last 15 years. While this is an increase, this rate is not appreciably different than the spent fuel transportation we have been experiencing. Spent fuel has been moving in the northeast. For example, in Connecticut spent nuclear fuel was transported by rail within the last month from Windsor Locks, through Hartford, CT, and Springfield, MA, to Idaho. In the past, spent fuel from Brookhaven Labs in New York was transported by barge across Long Island Sound through Connecticut. In Rhode Island in the last few months a shipment of spent nuclear fuel was made from the reactor at University of Rhode Island, through Connecticut and Pennsylvania, to storage at Savannah River, SC. In Maine, submarine fuel has been regularly transported west from the Kittery Naval Shipyard. Within the last two years in Massachusetts, spent fuel was shipped from the MIT reactor to Savannah River through Connecticut, New York and Pennsylvania. New York and Pennsylvania coordinated in 1993 and 1994 in the transportation of Shoreham Nuclear Plant spent fuel to the Limerick Nuclear Plant. In this same time period, in New Jersey spent fuel from Hope Creek Nuclear Plant was shipped through Delaware, Maryland, West Virginia and Pennsylvania on its way west to the General Electric Facility in Vallecitos, CA. The necessary planning, training and coordination has been accomplished for this spent fuel transportation and public safety has been protected. Moving spent fuel is not new in the Northeast.

Our Task Force is also aware of the good work accomplished by the Environmental and Operational Activities Division of DOE's Office of Civilian Radioactive Waste Management. Under this program, the Commercial Vehicle Safety Alliance has developed the Enhanced North American Standard Inspection for Transportation of Radioactive Materials for spent fuel transportation vehicles and drivers. This procedure was tested with impressive results in the Cesium shipment program from Colorado to Washington state in 1994. Transportation routing, interstate coordination and satellite tracking was tested in the Radioactive Nitric Acid shipment campaign from Hanford, Washington to Portsmouth, Virginia this year. Extensive transportation planning has been developed and training has been accomplished in the West in conjunction with the New Mexico Waste Isolation Project for high-level radioactive waste.

Finally, because we are the people involved in all types of material movement, we know that spent nuclear fuel transportation is less risky than much of the hazardous material now routinely transported. So we make the statement, transportation should not be an impediment for establishing interim storage.

VERMONT DEPARTMENT OF PUBLIC SERVICE

Now, shifting to the perspective of the state of Vermont and the Vermont Department of Public Service, let me say that I believe our views represent, to a great extent, the views of the other 38 states whose electric consumers pay for nuclear electricity and the waste program. We feel a great frustration with federal progress and policy in the spent nuclear fuel area. Since 1982, the $11 billion collected from electric consumers (over $1 billion from New England electric ratepayers) has resulted in $4.5 billion spent with negligible progress and $6.5 billion used for deficit reduction scoring. In this coming year, $590 million will be collected from consumers; DOE says it needs $600 million to meet its schedule; the Energy and Water Appropriation is $151 million from this source; and $439 million of ratepayer money is scored for budget reduction. Without interim storage, the frustration has reached such a level that the National Association of Regulatory Utility Commissioners (NARUC) is encouraging state investigations to withhold these funds from the federal government (Attachment B). Because of this frustration, we support S. 1271 which instructs the Department of Energy to design, construct and operate a facility for interim storage of spent nuclear fuel at a designated site.[1] The need for this legislation can be characterized in terms of three responsibilities: Monetary Responsibility, Environmental Responsibility and Governmental Responsibility.

[1] There is confidence that a proper plan exists for a spent fuel repository. However, based on past performance and present appropriations, there is no confidence in the completion schedule, presently stated as 2010, for the spent fuel repository. Spent fuel will have to be in temporary storage a long time, even after many nuclear plants have ceased operation. A single interim storage area will allow careful resolution of repository concerns without the pressure for hasty decisions.

MONETARY RESPONSIBILITY

First, in terms of monetary responsibility, electric consumers have paid large sums of money with precious little return as I have outlined above. This is compounded by the fact that consumers are paying two and three times again for the lack of program performance. Nuclear power stations were designed with limited onsite fuel storage in water filled pools and consumers have paid for that. Then consumers have paid to expand the storage capacity of spent fuel storage pools. Now, consumers are paying to move fuel from pools into dry cask storage. Finally in the limit, without a single interim storage facility, consumers will be forced to pay long-term custodial care for dry cask storage at 73 sites in 34 states, even after the nuclear plants themselves cease to exist.[2] Long-term care at power stations is estimated to be $5 billion more expensive than storage in a single interim storage facility. Under the prudence standard used by public utility commissions, costs for the waste program, either wholly or in part, would be candidates for findings of imprudence and excluded from ratepayer bills, were they not mandated by federal legislation. This proposed legislation, S. 1271, should be enacted on basic fairness principles for electric consumers.

ENVIRONMENTAL RESPONSIBILITY

The second responsibility is environmental responsibility. Without designation of an interim storage facility, spent fuel will remain on nuclear sites for a long, long time. One thing we know about this—the 73 nuclear sites would never be chosen in a site selection process for long-term storage of spent fuel because they are all, by necessity, adjacent to rivers or the ocean for cooling water. If dry cask storage were put at our nuclear plant in Vermont, Vermont Yankee, the dry casks would be exposed to the extremes of precipitation and freeze and thaws of Vermont winters and would be placed on the banks of the Connecticut River which runs directly through populated areas of Massachusetts and Connecticut to the Long Island Sound. We also know this—there are sites in dry climates, not adjacent to water bodies, which are well-suited for interim storage. For example, over $4.5 billion has been spent, much in the Nevada economy, to study the Nevada Nuclear Test Site. With all this expenditure, we do not know if the site is suitable for permanent disposal of spent fuel. But we know this about the Nevada Nuclear Test Site—it is nearly ideal for interim storage.

Another prime environmental consideration is the management of interim storage. As time goes on, there would be variations in the quality of management of up to 73 dry cask storage sites.[3] The chance for adverse environmental consequences will be worse at poorly managed sites. A single interim storage facility will allow focused attention on quality management of the site.

S. 1271, which designates a single site for the interim storage facility, is clearly the "green" preference to long-term storage at the 73 nuclear plant sites.

GOVERNMENTAL RESPONSIBILITY

The third responsibility is governmental responsibility. In 1958, our Vermont Department expressed concern about disposal of radioactive waste from the then proposed Yankee Rowe Nuclear Plant in western Massachusetts (Attachment C). The general public expected that government and industry would take care of this issue. In those days, the Congress was actively promoting nuclear power under the Atoms for Peace program. In the Declaration of Policy for the Atomic Energy Act of 1946 (Attachment D), the 79th Congress stated: "It is hereby declared to be the policy of the people of the United States that . . . the development and utilization of atomic energy shall, so far as practicable, be directed toward improving the public welfare, increasing the standard of living, strengthening free competition in private enterprise, and promoting world peace."

---

[2] Without the bill, the Yankee Rowe plant in Massachusetts, which permanently ceased operation in 1992, will have to collect millions of dollars more from New England electric consumers to pay for keeping its site open just to store spent fuel. With an interim storage site designated, and provision in the legislation to accelerate removal of waste from presently shutdown reactors, decommissioning of Yankee Rowe, Trojan and San Onofre may be completed in a timely, cost-effective manner. As time goes on, more and more nuclear sites will be in this condition.

[3] The Nuclear Regulatory Commission expends much of its resources enforcing the quality of nuclear operation. Even with this enforcement, it maintains a "problem plant list" of plants with performance problems. With the possibility of 73 dry cask interim storage sites, the prospect of a "problem site list" is real, not to mention the elaborate and expensive organization that will be required to watch over these sites.

The Congress made an implicit contract with the towns and localities which hosted nuclear facilities that spent fuel disposal would be solved. Since then, it has become clear that spent fuel disposal is far more difficult than the 79th Congress knew, and the federal promotion of nuclear power without this clearly worked out in advance may have been ill advised. Still, the federal responsibility over the spent fuel exists. We are now at a crossroads where more and more sites are building dry cask storage. Once all the sites have dry cask storage, and then all the reactors cease operation, it will be extremely difficult to muster the will to move the waste. These sites will become what has been labelled "Nuclear Stonehenges," memorials to a mistake in governing and to a lack of political will. The alternative is to establish one interim storage facility at the location most likely to be the disposal site, and concurrently to press ahead with the plan to evaluate and develop disposal. The presence of the interim storage facility can be expected to provide the impetus to complete the plan for disposal. Governmental responsibility dictates that S. 1271 should be enacted.

SUMMARY

Mr. Chairman, I know the spent fuel storage issue is difficult and contentious. Designating a location for an unpopular activity is always unpleasant. But the Congress makes all types of difficult decisions all the time. This issue is no different.

Transportation is not an impediment for interim storage. Responsible action in the monetary and environmental areas dictates establishing a single interim storage area. The country looks to the Congress for responsible government and responsibility in government dictates that government keep its promise to take care of nuclear waste. I urge the Committee, the Congress and the Administration to enact S. 1271.

Again, I thank you for the opportunity to make this statement today.

The CHAIRMAN. Thank you very much. We appreciate your statement

Mr. Snell.

STATEMENT OF DICK SNELL, CHAIRMAN, PRESIDENT AND CEO, PINNACLE WEST CAPITAL CORPORATION

Mr. SNELL. Thank you, Mr. Chairman. We most certainly do commend to this committee the bill introduced by Senator Craig. The Palo Verde operation that I represent in Arizona is not only the largest, but is also one of the newest units to come online, and in consequence we are most interested in the provisions of the bill relating to the interim storage and to the time table for building it. And in that respect we are particularly interested in those provisions of the bill that talk about 100,000-ton capability and retention of the current millage fee. And so by all means, we do commend this legislation.

[The prepared statement of Mr. Snell follows:]

PREPARED STATEMENT OF DICK SNELL, CHAIRMAN, PRESIDENT AND CEO, PINNACLE WEST CAPITAL CORPORATION

Mr. Chairman, members of the committee, my name is Richard Snell. I am chairman, president and chief executive officer of Pinnacle West Capital Corporation. Pinnacle West is the holding company for Arizona Public Service, which is the largest owner and operator of the Palo Verde nuclear power station. I would like to express my appreciation to the committee for inviting me this morning to present my company's perspective on S. 1271.

I want to begin by expressing my appreciation to you, Mr. Chairman, to the ranking minority member, Mr. Johnston, and to the other members of this committee for your interest in the U.S. high-level waste program, and your commitment to set this program back on course. I would like to extend special thanks to Mr. Craig and Mr. Kyl and the other members of the Senate who have co-sponsored S. 1271, the Nuclear Waste Policy Act of 1995.

THE ISSUE IN BRIEF

Mr. Chairman, you heard testimony in March from Sam Skinner, president of Commonwealth Edison Co., on behalf of the Nuclear Energy Institute about the need for this nation to take a systematic approach to spent nuclear fuel management. We support the framework that Mr. Skinner laid out, including the need for the federal government to create an organized, integrated system that includes:

- a federal central storage facility to allow for the centralized management of spent fuel until a federal permanent disposal facility is ready;
- a permanent disposal facility; and
- a transportation system to move spent fuel safely from nuclear power plants to the storage and disposal facilities.

Such an integrated system should be designed to enable the federal government to meet its responsibility to accept spent nuclear fuel starting in 1998. This was the date specified when the Nuclear Waste Policy Act was passed in 1982, and affirmed by the U.S. Department of Energy several times since then. Stated simply, the federal government has a clear responsibility to begin accepting spent nuclear fuel in 1998. Electricity consumers have held up their side of the deal. The federal government must do the same.

HISTORY OF THE SPENT FUEL PROGRAM

The committee is well aware of the history of the spent fuel management program and I don't want to repeat that here today. However, there are a few points that are important to make in the framework of my testimony.

This nation's program to manage spent nuclear fuel is a chronicle of missed milestones and broken promises. In 1982, Congress passed the Nuclear Waste Policy Act, which assigned responsibility for management of spent nuclear fuel from our nation's nuclear power plants to the U.S. Department of Energy. Congress was told that permanent disposal facilities would be ready in 1998, and that date was written into the law.

The legislation also created a Nuclear Waste Fund to cover the cost of this program. Since then, consumers of electricity from nuclear power plants, pursuant to contracts signed between the Department of Energy and U.S. nuclear utilities, have paid a surcharge of one-tenth of a cent per kilowatt- hour into this Fund. To date, utility customers have committed nearly $12 billion for spent fuel management. Customers who use electricity generated at Palo Verde have paid $167.8 million for these federal government services. By the time the three Palo Verde reactors reach the end of their initial 40-year license, electricity consumers will have contributed nearly $1 billion to the Nuclear Waste Fund.

Mr. Chairman, this deal was not open to negotiation. Utilities were required to sign contracts with the Department of Energy for management of this fuel in order to continue to operate their nuclear power plants.

In 1987, Congress amended the Nuclear Waste Policy Act and designated Yucca Mountain, Nevada, as the only repository site to be initially evaluated. The 1987 amendments also implicitly acknowledged what many suspected: that a permanent disposal facility would not be ready by 1998.

While attempts to implement the Nuclear Waste Policy Act of 1982, and that act's 1987 amendments, have begun to address the technical challenges of licensing a repository, we have yet to fully address the complex political dynamics that surround this issue.

Mr. Chairman, the expectations among the states, utility regulators and the utilities themselves are very clear. They expect the federal government to honor its commitment to begin managing spent nuclear fuel in 1998.

Last year, utilities and agencies in 34 states filed similar, but separate, lawsuits against the Department of Energy—the most dramatic evidence to date of the growing frustration over the Department of Energy's reluctance to accept its responsibility. Those lawsuits asked the D.C. Circuit Court of Appeals to rule that DOE must begin to accept spent nuclear fuel beginning in 1998. This litigation is still pending.

The nuclear energy industry would like to acknowledge the priority attention devoted to this program by Secretary O'Leary and her staff. Under Secretary O'Leary's leadership, efforts relative to the Yucca Mountain project have been redoubled in order to maintain the schedule. Site characterization has progressed in earnest. Yet, uncertainty continues to plague the program and there is no clear schedule of when the energy agency expects to begin accepting spent fuel at a repository.

BROAD SUPPORT FOR CORRECTIVE LEGISLATION

Unlike many public policy matters, spent nuclear fuel management is one of those rare issues on which there is an emerging broad consensus. Many stakeholder groups—including state regulators, state legislators, state attorneys general, the nuclear energy industry, labor unions and the American people—agree on the nature of the Energy Department's responsibility and on what must be done with this program to set it on a course that will ensure success.

As was discussed at your March hearing, while the consensus remains that geologic disposal of spent nuclear fuel in an underground repository is the safest means of permanent disposal, the process or system that can lead to that end point has never been clearly articulated. The ultimate goal—deep geologic disposal—remains valid today, but this nation must establish an integrated management system to reach that destination. We at Arizona Public Service believe there is an urgent need for a central storage facility that can safely and efficiently manage high-level radioactive waste until a permanent solution is available.

An integrated system has three essential components: a permanent disposal facility; a central storage facility until the permanent facility is ready; and a transportation infrastructure to move spent fuel from reactor sites to the storage and disposal facilities. Remove one of those components, and the safety and efficiency benefits of integration are lost.

Unfortunately, that is where the United States stands today because the Department of Energy does not have a central storage facility, is not planning to build one and—because of restrictions placed on it by current law—does not have the authority to build such a facility.

Clearly, the Department of Energy will be unable to meet its 1998 responsibility absent a significant shift in program direction. Despite continuing efforts, the nuclear energy industry and state regulators have been unable to fix the problem. Only Congress can take the appropriate steps to set this program back in the right direction, because only Congress represents all those with a stake in this issue and only Congress can create the legal framework and provide the necessary oversight to get the job done.

The last time this committee examined possible legislation to develop an integrated system for managing used nuclear fuel, the only potential vehicle in Congress was a draft bill, which has since been reported out of the Commerce Committee and has strong bipartisan support in the House.

I would like to commend Senators Craig and Kyl for introducing a vehicle in the Senate that seeks similar reforms in the civilian waste disposal program and to ensure that the nation safely stores byproducts from atomic energy defense activities and research reactors. S. 1271 incorporates a number of principles and provisions that are essential to setting this program on course. Among those provisions, Senator Craig's bill:

- Reaffirms the federal government's responsibility to accept, and take title to, spent fuel from nuclear power plants for storage and disposal starting no later than January 31, 1998.
- Designates a location for—and mandates construction of—the central storage facility that is essential if the Department of Energy is to meet that 1998 responsibility and eliminates the linkage in current law between licensing of a repository and construction of a central storage facility.
- Provides the necessary transportation corridors and rights-of-way, and authorizes construction of any necessary rail spurs, to ensure access to that central storage facility from existing rail lines.
- Establishes explicit milestones and expedited schedules for preparation of environmental documents, conduct of licensing reviews and all other steps involved in siting, design, licensing and construction of this central storage facility.
- Ensures that the scientific characterization of the Yucca Mountain, Nevada, site will continue, although flexibility is provided in the milestones consistent with available appropriations. Development of the central storage facility takes first priority, followed by (2) the transportation infrastructure and (3) repository development.
- Ensures that the Nuclear Waste Fund fee remains at one mill per kilowatt-hour.

Senator Craig's legislation represents important movement in the Senate in the growing effort to make the federal government accountable for honoring its contracts to accept spent fuel. S. 1271 modifies the House approach to include these important provisions:

- The central storage facility has a maximum capacity of 20,000 metric tons of uranium in the first phase and an initial 20-year license, which is renewable. The second phase of the facility, which must begin operation by Dec. 31, 2002, has a capacity of 100,000 metric tons of uranium and an initial licensing term of up to 100 years;
- Grants DOE authority to store additional spent fuel and high-level radioactive waste, such as Naval spent fuel and used fuel from shut down reactors and re-

search reactors, once the Energy Secretary has achieved the annual acceptance rate from civilian reactors;

- Retains the current 1-mill per kilowatt-hour fee to fund the program and caps the fee at that level, and
- Provides flexibility in the repository schedule due to uncertain funding.

Mr. Chairman, an important difference in Senator Craig's bill is the development of a central storage facility that is large enough to accommodate all of the utilities that have contracts with the Energy Department, as well as high-level radioactive waste from atomic energy defense activities.

In October, Congress passed the Energy and Water Development Appropriations Act that included just $400 million for the Department of Energy's civilian waste management program in fiscal year 1996. Despite the fact that nuclear utility consumers paid approximately $600 million into the Nuclear Waste Fund in 1995, Congress allocated just $151.6 million from the trust fund for work at Yucca Mountain in this fiscal year. As a result of funding levels that are dramatically lower than expected by the Department of Energy, the agency is reexamining its program objectives and whether it can meet current milestones at Yucca Mountain. Given the history of this program, one can only anticipate yet another delay at Yucca Mountain—which makes the development of an expanded central storage facility all the more critical.

Furthermore, we can support the current funding situation if the program is changed to give electric utility consumers the essence of the deal they made with the federal government—spent fuel acceptance beginning in 1998 in exchange for program funding.

In the event that the Office of Civilian Radioactive Waste Management activities continue to be constrained by funding shortfalls, the industry believes that provisions in S. 1271 to prioritize DOE's waste management components are appropriate. In preparing annual budget requests, the Energy Secretary must make development and operation of a central storage facility the priority, followed by development of a transportation program and, finally, continued site characterization of Yucca Mountain.

Mr. Chairman, this ranking does not reflect a sudden lack of need for a repository. A deep geologic repository, located at an isolated, arid environment, is still the cornerstone of the nation's spent fuel management policy. It is an essential facility not only for commercial nuclear fuel, but for the high-level radioactive waste from defense and other national programs that is presently stored in about 30 states. Locating a central storage facility near Yucca Mountain is the sensible approach to an integrated system and will ease transportation of used nuclear fuel to the repository.

Finally, Mr. Chairman, I think it is essential for Congress to provide adequate benefits for the host state and local governments willing to host an element of the integrated spent fuel management system. The nuclear industry has consistently stated the view that a cooperative relationship between all governmental entities is preferable, if not essential, to the development of a waste management system. We urge this Committee to fully explore this issue and hope that you will be able to support the establishment of an effective benefits provision—one that both recognizes the various equities and addresses state or local impediments, real or perceived, to the achievement of what is fair.

THE CONSEQUENCES OF NO NEW LEGISLATION

Based on repeated promises from the federal government, electric utilities that built nuclear power plants expected spent fuel to be stored for a short period of time on site, then shipped off-site either for reprocessing or storage and disposal at federal government facilities. On-site spent fuel storage facilities were never intended to provide life-of-plant storage capacity. Unless the Department of Energy moves quickly to build a central storage facility, Arizona Public Service and our customers will face serious negative economic impacts. Our three-unit Palo Verde station can operate until about 2003 without additional storage capacity. Beyond that point, however, we would be forced to plan additional life-of-plant storage capacity at a cost of between $150–$175 million. This, of course, is in addition to nearly $1 billion that consumers of electricity from Palo Verde will pay into the Nuclear Waste Fund over the 40-year life of the three units.

Continued delays in the Energy Department's program to accept spent fuel will subject all utility customers to significant added costs for additional central storage capacity at plant sites.

We note that a number of utilities have already been forced to build additional on-site storage capacity. In some cases, this has subjected those companies to unacceptably high political and financial costs. Precipitating these battles on a state-by-state basis is not what federal policy intended. It is not fair to expose utility customers and state legislatures to this kind of highly charged conflict, and it is avoidable. As difficult as they are today, these issues will only get more complex.

CONCLUSION

The facts before this committee and this Congress are clear and simple: This nation has 109 nuclear power units operating, providing more than 20 percent of our electricity, and producing spent fuel. It is up to us, who enjoy the benefits of that electricity, to manage the waste byproducts safely and responsibly. We cannot simply drop this problem in our children's laps.

S. 1271 will correct problems that have been evolving for more than 10 years. The schedules and milestones necessary to site, build and license a central storage facility by 1998 are tight, and do not allow for slippage, delay or further excuses.

We in the nuclear energy industry fully expect that some will try to persuade the members of this committee and this Congress that the job cannot be done in the time allowed, and that the schedule is too demanding. In our view, however, all those who have a stake in this issue have a right and a responsibility to be demanding.

Likewise, Arizona Public Service and other electric utilities have a legitimate right to expect performance—to safeguard the billions of dollars invested in nuclear power plants by our customers and shareholders.

Without a central storage facility and a means of satisfying the existing federal responsibility to accept spent fuel starting in 1998—more and more nuclear power plants will run out of on-site storage space for spent fuel. By 1998, 26 nuclear units will have exhausted their pool storage capacity. By 2010—the earliest date for opening a permanent disposal facility—80 units will be out of such space.

Again, I would like to commend Senator Craig for his legislation and I encourage the committee to markup this legislation as soon as possible.

The CHAIRMAN. Would you care to comment very briefly about the capacity of your plant relative to MO_x fuel and what the current authority might be, if you had it? Have you ever tested your plant with it?

Mr. SNELL. No, we have not, Senator. I am told that technically there is no question it could be done. It would require an amendment to the license from the NRC. Politically, we would have a major problem not only nationally but in the State of Arizona. And I would defer to you in that respect.

The CHAIRMAN. Well, that is fair enough. If you wanted to test your plant, I assume you could get MO_x fuel from France and run a test. Would it require significant alteration of your plant, or do you just put it in as a run of the mill energy source?

Mr. SNELL. My understanding is that some change would have to be made in the fuel rods. There is the licensing issue, but it could be done.

The CHAIRMAN. I guess my point is that within the industry and the current technical capability, there is an ability to basically burn the waste as opposed to store the waste.

Mr. SNELL. I cannot speak for plants other than my own, but my understanding again is with alteration of the control rods it could be done.

The CHAIRMAN. Why is yours unique? Was it designed with potentially this in mind as an alternative?

Mr. SNELL. My understanding is that combustion engineering, which is the source of the design for our plant, did have it in mind. And there are a few other plants around the country, I am told, that can do it. Also, the can-do plants in Canada.

The CHAIRMAN. Now, Canada is considering doing some testing up there. Are any of you knowledgeable of some requests, either by the DOE or others, to watch that Canadian effort? I assume the Canadian plants are not experimenting with MO_x.

Mr. SNELL. I cannot answer in terms of formal request. I do know that discussions have been held.

The CHAIRMAN. I think that there is some discussion about the Canadians burning both Russian plutonium and some of our defense plutonium. And I do not want to prolong this because it is late, but I think we should be aware that we are not exactly neophyte in this process as we look into the next age of technology, which hopefully will give us some relief from the storage problems we have.

Mr. McNeill.

STATEMENT OF CORBIN A. McNEILL, JR., PRESIDENT AND CEO, PECO ENERGY COMPANY

Mr. MCNEILL. Thank you very much, Mr. Chairman. I will be brief here.

For many people transportation has emerged as a central issue in the debate over the development of an integrated spent fuel management program, and I just want to briefly review some of PECO Energy's recent experience, which is very germane here.

In 1993, my company, PECO Energy Company, agreed with the Long Island Power Authority, then known as the Shoreham Nuclear Powerplant. We agreed to accept the slightly irradiated fuel from that plant which was being decommissioned, and we accepted that fuel for our Limerick generating station, which is about 35 miles northwest of Philadelphia. This material, the Shoreham fuel, was shipped as if it were fully spent nuclear fuel, utilizing the same transportation casks and the same regulatory programs required for moving spent fuel.

In an effort to educate the public and to counter disinformation being disseminated about the transportation of nuclear fuel, we produced a short videotape which recounted the history of nuclear fuel shipments and showed the dramatic crash tests of similar containers which were tested at the Sandia National Lab in New Mexico. The footage there offers dramatic proof of the strength and integrity of these robust containers, and I would like each of you to view the video if you have not done so.

After extensive media coverage of our first shipment, we successfully transported 33 casks with 200 tons of fuel without a significant or even a single incident, and after the first shipment without any additional media attention, PECO Energy's experience with Shoreham fuel shipments provided the company and the entire industry with insights about the shipment of nuclear fuel. The transportation of nuclear fuel can be accomplished without public concern as long as the media, State and local officials, and the public are well educated. As the number of shipments of radioactive material increase, the industry and the Federal Government will have to increase their efforts to inform and educate these groups.

Let me close by encouraging you and the members of your committee to move quickly and without further delay to mark up S. 1271, and report it to the Senate floor.

[The prepared statement of Mr. McNeill follows:]

PREPARED STATEMENT OF CORBIN A. MCNEILL, JR., PRESIDENT AND CEO, PECO ENERGY COMPANY

Mr. Chairman, members of the committee, my name is Corbin McNeill. I am President and Chief Executive Officer of PECO Energy Company. PECO owns all or part of six nuclear power plants in Pennsylvania and New Jersey, and we operate four of those units. Nuclear energy generates approximately 60 percent of the electricity that PECO Energy requires to serve 1.5 million customers in Philadelphia and southeastern Pennsylvania.

On behalf of the nuclear energy industry, I would like to express our appreciation to you, Senator Murkowski, to the ranking minority member, Senator Johnston, and to the other members of this committee for your interest in the U.S. commercial high-level waste program. We commend you for your commitment, as evidenced during this series of hearings, to set this program back on course. I would also like to acknowledge the leadership of Senator Craig and Senator Kyl for their efforts in introducing Senate Bill 1271, which provides the reforms necessary to set the government's high-level waste program back on course.

SUPPORT FOR S. 1271

I cannot overemphasize the nuclear energy industry's strong support for this legislation, which is now cosponsored by 20 members of the Senate. We look forward to working with this Committee to move this legislation quickly through committee to the Senate floor. This bill, similar in many respects to H.R. 1020, which has the support of a bipartisan majority of the House of Representatives, provides a solid foundation on which to proceed.

We believe that the U.S. government and the commercial nuclear energy industry share a common goal. We must create an organized, integrated system to manage spent nuclear fuel that includes:

- a federal interim storage facility to allow for the centralized management of spent fuel until a federal permanent disposal facility is ready;
- a permanent disposal facility; and
- a transportation system.

When it is enacted, S. 1271 will enable the federal government to meet its commitment to accept spent nuclear fuel starting in 1998. This was the date specified when the Nuclear Waste Policy Act was passed in 1982, and affirmed by the Department of Energy many times in the 13 years since.

TRANSPORTATION OF SPENT FUEL

I would like to focus the balance of my testimony on PECO's experience with the transportation of nuclear fuel, because my company has unique experience in this area, having transported slightly irradiated fuel from the Shoreham nuclear power plant on Long Island to our Limerick station.

For many people, transportation has emerged as a central issue in the debate over the development of an integrated spent fuel management system. Before I discuss in detail the events surrounding the shipment of fuel from the decommissioned Shoreham nuclear plant, let me provide some general background on transportation of spent nuclear fuel.

The nuclear energy industry has an established transportation program that has shipped more than 3 million packages of radioactive materials—including low-level and high-level radioactive waste—since 1964. Of the 100 million packages of hazardous materials shipped each year, 97 percent are flammable, explosive, corrosive or poisonous. Radioactive waste accounts for just 3 million packages, or 3 percent, of the hazardous waste shipped in this country.

High-level radioactive waste, including spent nuclear fuel, has been shipped from nuclear power plants to facilities that were to reprocess used fuel. Spent fuel from a now-closed reprocessing facility in West Valley, N.Y., has been shipped back to utilities and used fuel at the Morris, Ill., facility remains in safe storage there. Damaged fuel from the Three Mile Island plant was shipped to the Idaho National Engineering Laboratory during the late 1980s, and a small number of spent fuel packages have been sent to the national laboratories for research and development activities.

High-level waste produced at nuclear power plants, most of it in the form of used nuclear fuel rods, amounts to about 2,000 tons each year. Most of this spent fuel is stored at nuclear power plants in steel-lined concrete pools or in rugged, air-cooled steel or concrete containers. Over the past 25 years, 2,388 shipments of spent nuclear fuel have been transported safely over the nation's highways, waterways and railroads.

Spent nuclear fuel consists of solid, ceramic material placed end-to-end in zirconium alloy tubes called fuel rods. The tubes are configured into fuel assemblies—typically 12 inches square and 12 feet long. After being used to fuel a nuclear power plant, the assemblies are cooled in pools at the plant site before being placed in dry, rugged containers for shipment. These specially designed containers—certified by the Nuclear Regulatory Commission—use heavy metal-walled technology to provide shielding and containment of the spent fuel.

Using existing container technology, the nuclear energy industry and others responsible for spent nuclear fuel shipments have established an exemplary safety record. No fatalities, injuries or environmental damage have occurred because of radioactivity of the cargo.

Shipments of spent nuclear fuel rods should begin increasing in 1998 when the federal government is responsible to begin accepting spent fuel from utilities. The Energy Department is responsible for transporting, storing and disposing of used nuclear fuel under provisions of the Nuclear Waste Policy Act. Under the law and contracts between utilities and DOE, the agency is responsible for accepting spent fuel from nuclear power plants for storage or disposal in 1998.

Because nuclear fuel is an extremely compact source of energy and because the volumes are small, the number of anticipated shipments of spent fuel to storage and disposal facilities is also quite small—averaging about 300 per year starting in 1998.

EXISTING REGULATIONS ENSURE SAFETY

The existing regulatory program for shipment of radioactive material ensures public health and safety as well as protection of the environment from dangerous releases of radioactivity. Several federal agencies play a role in the safe transportation of radioactive materials:

- The Department of Energy is responsible for accepting, transporting, storing and disposing of spent fuel from nuclear power plants, which generate 20 percent of America's electricity.
- The Department of Transportation regulates highway routing, packaging, labeling, shipping papers, personnel training, loading and unloading, handling and storage as well as transportation vehicle requirements.
- The Nuclear Regulatory Commission must approve shipping container design and manufacturing to ensure they maintain their integrity under routine transportation conditions and during severe accidents. The agency also examines shipping routes to ensure the security of spent fuel shipments.

The Department of Transportation regulates all shippers and carriers of radioactive materials and requires that they comply with all of the agencies' regulations. Before radioactive material is transported, DOT requires the shipper to contact the governor of any state through which the cargo will pass. The governor may take additional public safety actions deemed necessary, such as providing escorts for the shipments.

Transportation Department regulations also require shippers to use the most direct routes for hazardous material shipments. States can also work with DOT to establish preferred highway routes and time periods for shipments. The public can be involved in this process at the state level. The transporter must comply with regulations that may stipulate a specific notification process, weight limits and time-of-day restrictions.

The Nuclear Regulatory Commission establishes safe standards for radiation limits during shipment. Transportation containers are designed to withstand accidents more serious than they will ever likely face—without releasing their contents.

Before the NRC certifies a container for transportation of spent nuclear fuel, the container must meet rigorous engineering and safety criteria and pass a sequence of hypothetical accident conditions that create forces greater than containers will experience in actual accidents, including:

- A 30-foot free fall onto an unyielding surface, which is equivalent to a crash into a concrete bridge abutment at about 120 miles per hour.
- A puncture test allowing the container to fall 40 inches onto a steel rod 6 inches in diameter.
- A 30-minute exposure at 1,475 degrees Fahrenheit that engulfs the entire container.
- Submergence of that same container under 3 feet of water for eight hours. Containers are also subject to separate testing under 50 feet of water for eight hours.

According to NRC regulations, the radiation level of containers during shipment cannot exceed 10 millirem per hour at a distance of about six feet from the truck. At this level, a person who spends 15 minutes standing six feet away from the vehicle carrying radioactive materials would receive 2.5 millirem of radiation. By comparison, the average person receives about 300 millirem each year from natural background radiation, including an average of 45 millirem from routine medical X-rays.

Between 1971 and 1989, just seven accidents have occurred involving transportation of spent nuclear fuel and none resulted in the release of radioactivity.

The most severe of the seven occurred in 1971 in Tennessee. A tractor-trailer carrying a 25-ton spent fuel shipping container swerved to avoid a head-on collision, went out of control and overturned. The trailer, with the container still attached, broke free of the tractor and skidded into a rain-filled ditch. The container suffered minor damage, but released no radioactive material.

Several factors have contributed to the excellent safety record for spent fuel shipments:

- Strict requirements on packaging and handling of radioactive materials, which are subject to stringent quality control by all parties, including the NRC, utilities and transportation companies.
- An established system for responding to accidents involving radioactive cargo.
- A voluntary mutual assistance agreement among nuclear utilities on responding to an accident.

LOCAL EMERGENCY RESPONSE TRAINING

S. 1271 would provide funding from the Nuclear Waste Fund to provide emergency response training for state and local entities in the event of an accident involving radioactive materials.

Using this funding, DOE provides assistance for training state and local officials in developing emergency response and preparedness plans, and training emergency response professionals in handling accident situations. The agency also operates a Radiological Assistance Program, with eight regional offices staffed with experts available for immediate assistance. If necessary, police will summon those experts to handle the transportation package and remove any radioactive material that may have been released.

The Transportation Department has also developed a self-contained training course on handling emergencies involving radioactive materials. In the event of an accident, the carrier has basic responsibility for confining the spread of radioactive materials and for any cleanup. State and local government agencies, through their police and fire departments, normally are responsible for protecting citizens and property at the scene of the accident.

PECO'S EXPERIENCE WITH NUCLEAR FUEL TRANSPORTATION

On March 1, 1993, PECO Energy Company agreed with the Long Island Power Authority (LIPA) to accept slightly irradiated fuel from the Shoreham nuclear power plant, which was being decommissioned. PECO Energy accepted the fuel for use in our Limerick Generating Station 35 miles northwest of Philadelphia.

Unlike fresh fuel, which can be handled without protective shielding, the Shoreham fuel shipped as if it were fully spent nuclear fuel, utilizing the same transportation casks and the same regulatory program required for spent fuel.

PECO Energy worked with the U.S. Department of Transportation, including the Coast Guard, and the Nuclear Regulatory Commission, the Pennsylvania State Police as well as state and local government officials, to obtain the necessary regulatory approvals and permits.

To minimize public concerns regarding the shipment of the fuel, PECO Energy and LIPA agreed to ship the fuel by barge from Long Island, down the Atlantic Coast, and up the Delaware River to PECO Energy's coal-fired Eddystone Generating Station just south of Philadelphia. Once the casks were offloaded and put onto a rail car, the shipment then proceeded to Limerick by rail during low-traffic periods.

In an effort to educate the public and counter disinformation being disseminated about the transportation of nuclear fuel, PECO Energy produced a short videotape recounting the history of nuclear fuel shipments and showing dramatic crash tests of similar containers being tested by Sandia National Lab in New Mexico. These tests, as I mentioned earlier, included running a jet-assisted flatbed tractor trailer with a container into a concrete wall at 80 miles per hour and another tractor trailer being broadsided by a jet-assisted train locomotive traveling 80 miles per hour. The footage offers rather dramatic proof of the strength and integrity of these enormous containers, and I would urge each of you to view the videotape if you have not done so already.

To give you a sense of proportion, the casks used to transport the fuel weighed 70 tons. Each cask contained 17 fuel assemblies weighing a total of 20 tons. Thus, for every ton of fuel, there were 3.5 tons of protective shielding.

In another effort to build public confidence, we invited three reporters, including a television camera crew, to accompany PECO Energy Chairman Joe Paquette on the initial train shipment from Eddystone to Limerick on September 27, 1993. The trip was uneventful, and media coverage and public interest quickly dissipated after the first shipment was completed. The final 32 shipments progressed without a single line of type in any newspaper and without a word of mention on either radio or television. The final fuel shipment arrived at Limerick nine months later on June 5, 1994.

In the end, we successfully transported 33 casks with 200 tons of fuel without a single incident.

PECO Energy's experience with the Shoreham fuel shipments taught the Company and the entire industry important lessons about the shipment of nuclear fuel. The transportation of nuclear fuel can be accomplished without public concern as long as the media, state and local officials, and the public are well-educated about the issue. As the number of shipments of radioactive material increase, the industry and the federal government will have to increase their efforts to inform and educate these groups.

The Shoreham fuel project provided PECO Energy with a "dress rehearsal" for the long awaited day that fuel will be shipped to a federal interim storage facility. The industry has proved that the transportation of spent nuclear fuel can be accomplished with complete protection of public health and safety and the environment. The real challenge is to effectively communicate the safety of this activity to the public.

CONCLUSION

Mr. Chairman, let me close by encouraging you and the members of this committee to move quickly and without further delay to mark up S. 1271 and report it to the Senate floor.

Although the political dynamics of the spent fuel issue are undeniably complex, the solution—which is articulated in S. 1271—is very simple. The federal government must meet its responsibility to move spent nuclear fuel starting in 1998. To do so, the government must build a central storage facility to hold that fuel until a permanent disposal facility is ready. The time for debate and study is long since over. Over the last 13 years, my industry's customers have committed over $11 billion for that service. You and I share a solemn responsibility to ensure that the job gets done.

Thank you.

The CHAIRMAN. Thank you, Mr. McNeill.

Mr. Sherman, you seem to be, among others, an expert on transporting your fuel. I am talking about primarily the waste. Recognizing the dilemma associated with access into Yucca and whether you build a billion dollar railroad, there is a shorter alternative, and that would be to use the existing railroad capability to go around north Las Vegas and come in the back into Yucca. Is there any reason, in your opinion, why that should have any significant exposure, from the standpoint of safety, to use an existing rail?

Mr. SHERMAN. Based upon all of the experience in transporting spent fuel, both in the military and in commercial spent fuel, it can be done safely on either route.

Mr. MCNEILL. Mr. Murkowski, if I might interject here, the fuel that we moved was shipped by rail through the city of Philadelphia itself on conventional conrail railroad tracks.

The CHAIRMAN. The only difference I can see is Philadelphia goes to bed at night and Las Vegas wakes up.

[Laughter.]

The CHAIRMAN. Maybe in Las Vegas you should ship it through in the daytime.

[Laughter.]

The CHAIRMAN. Did you have any further remarks?

Mr. MCNEILL. No, I fully agree with Mr. Sherman that there is no reason.

The CHAIRMAN. And in Europe they ship all the time by rail, right?

Mr. MCNEILL. That is correct. And in fact, all of our Navy fuel is shipped by rail.

The CHAIRMAN. Can you gentlemen give us any comfort on how the American public might be reassured that indeed this process is safe and the sense that the risk can be reduced to a point where it is no more risky than the sky falling in?

Mr. MCNEILL. Well, we had, in the Philadelphia area, we initially had local political opposition. It was not strong. We, in an extensive campaign of informing the public, meeting with the public before we did any shipments, we developed this video, and we worked with State and local agencies to reassure them, and actually in meeting with the Philadelphia Inquirer Editorial Board we made the point that Mr. Sherman has made, that the risk associated with this shipment is substantially less than other hazardous shipments that occur in this country on a routine basis. And I was very favorably impressed when they put that into an op ed piece that they did, that they made that comparison for us.

The CHAIRMAN. Do you still have a copy of that?

Mr. MCNEILL. I can provide it to you.

Mr. SHERMAN. And, Mr. Chairman, if I may speak to that question, as well, one place where the Department of Energy is doing good work, and I cover it in my written testimony, is in the area of cooperative groups where they organize State groups such as the transportation task force that I represent. In that we have studied and interacted and had an influence on transportation considerations. I believe that these groups are going a long way to help do the things that will make public opinion accept the transportation of spent fuel.

The CHAIRMAN. And that is applicable to Nevada, too, and the delegation. Mr. Snell, did you have a comment?

Mr. SNELL. I know the Secretary testified to a lack of public acceptance. But I would argue that in fact there is a good deal more public acceptance than she talks about. I think there is a consensus among State regulators and attorneys general and Governors and the industry and the American people, as a matter of fact, that we really need to get on with this.

The CHAIRMAN. I am going to call on Senator Johnston, and then I am going to pass the gavel to Senator Craig. I am going to have to excuse myself. But I do want to compliment both you, Senator Craig, and Senator Johnston for your effort on the bill before us. I think the testimony has been worthwhile, and with the Secretary of Energy's willingness to explore some avenues that previously had been off limits have now been opened up to an opportunity to develop a dialogue which may expedite the process.

And again, as I indicated in my opening statement, I think we all agree we want to do it right this time. And that is going to be basically my bottom line as I attempt to develop a chairman's mark on this as we proceed with hopefully a final disposition of this. And I want to thank you for being here.

Senator CRAIG. Mr. Chairman, thank you.

The CHAIRMAN. I am going to call on Bennett.

Senator JOHNSTON. Thank you, Mr. Chairman.

Gentlemen, just an observation. I would submit to you that transportation of nuclear waste is orders of magnitude safer than routine shipments of chemicals, orders of magnitude, there has never been an accident. If there is an accident, you have got a break-proof container that you can drop off buildings, you can hit with trains, and it does not break open, and even if it did, the likelihood of there really being a catastrophic accident is fairly slight, because it does not blow up, it does not flow into the water supply, it can be cleaned up—et cetera, et cetera. Now, whether you can convince the public of that is another question, but it is clearly scientifically true and demonstrable.

We just had an accident down in my State just last month on chemical waste. We have them all the time. It is much more dangerous, orders of magnitude more dangerous, than shipment of nuclear waste.

When we do get to markup on this bill, and I hope we do soon, and I would predict that Senator Craig and I and Senator Murkowski can work out differences because we have no different ends in mind, we want to solve the problem. I hope you will help on this question of the 1998 date. Now, the industry has been strong for fixing a 1998 date. Why is that? Not because 1998 is a date that anybody has ever thought was a achievable for interim or any kind of storage. It cannot be. But for the purpose of fixing legal liability on the part of the Federal Government, which is fine, and you can make a very strong case that they ought to be liable, they have been at fault, they have delayed, et cetera.

But I have yet to have any lawyer on any side of this controversy tell me that you can avoid the point of order that liability on behalf of the Government would create. That is, it creates a liability, probably to be paid out of what we call the judgment fund, which in my view creates a point of order, which would mean that you then have to have 60 votes to waive the Budget Act, which I do not believe that you could get done. Now, if anybody would suggest to me something different—there is some way to get around that and to have it paid out of the appropriations, out of the waste fund, which would mean it would delay that much more.

First of all, it would be difficult to get the appropriation specifically for that purpose; and secondly, it would delay by that much your ability to get money to build your interim facility. And I do not believe the industry wants to fund out of the one mill, out of the monies appropriated, the building of dry cask storage at those reactors that need it. Now, maybe they do. I do not believe so. But it is a vain hope, in my view, to say that you ought to have a 1998 date creating legal liability that the Government is going to fund, and that you can get that passed in this bill. And what it does, it creates your point of order and you cannot pass the bill.

So I hope you will take that message to Garcia back to the industry. My purpose is not to deprive you of your ability to sue the Government, but we need to get the legislation passed. And there is a serious point of order. I keep raising it every time somebody comes by the office. Nobody ever has, nobody ever argues with it, nobody ever disagrees with it, but we keep getting a very strong industry response that says we have got to have that, do or die. Now, I say let us get legislation passed. It is important to get it underway and not have it subject to a point of order.

Secondly, we need to be thinking about questions like the route. You know, there are three different routes here, one that goes basically on the outskirts of Las Vegas, one that goes through Nellis Air Force Base, and one that goes way around, the billion dollar route. Now, you cannot have everything. You cannot put limits on the 1 mill per kilowatt hour and then also pay Nevada all kind of monies and build an interim facility and have it all come out even. Somewhere you have got to make two plus two equal four. So the industry needs to give some thought on those kinds of questions, because this is a cooperative effort to pass a bill that can both be passed and funded to pay for a repository and an interim facility.

You have made a very strong case. The problem is not making a strong case, it is getting Congress off the dime. I happen to believe we can override a veto on this bill. Look, the waste is there. It has got to be dealt with. Now, if the environmental community thinks that they are making progress by killing legislation dealing with repositories, they have not thought through the fact that you have got 34 States that have this stuff, like you, Mr. Sherman, said, in Vermont, that drains right into Long Island Sound. I mean, these are serious problems, and we need to deal with them in a serious, scientific way, safely.

I am totally convinced this can be done totally safely and it is not an engineering challenge that mystifies any of the engineers or a scientific challenge. They have known the science, they have known the engineering for a long time. It is the politics and the way we wrap ourselves around the axle on impossible requirements sometimes, both political and other.

So all of that by way of saying I thank you for your testimony and I confidently predict that if we can get to the markup on this bill that Senator Craig and I and Senator Murkowski and others can put together good legislation. But keep in mind this 1998 date. If you can get your lawyers to come in and convince me that you can put in a 1998 date, create legal liability on the Federal Government, and not have any budget problems, then count me in. But I have been on the budget committee a long time, and I think I know the problem which that presents. And what I want to do is get a bill as quickly as we can.

Thank you, gentlemen, and thank you, Mr. Chairman.

Senator CRAIG [presiding]. Bennett, let me thank you. And what I think is important for the committee record is to demonstrate the resolve that you have and that I have and that I think the chairman has to deal with this issue, to get a markup vehicle, and to get it to the floor of the Senate for a vote, and to work out our differences with the House.

Now, in that regard, Mr. Gates, let me suggest that with the urgency of your testimony that you talk to your Senators and ask them to become cosponsors of this fine legislation.

Mr. GATES. I will.

Senator CRAIG. Thank you. That would be very helpful.

Senator JOHNSTON. Maybe we should suggest that to all the panel members.

Senator CRAIG. You all keep working on it, then.

Mr. Sherman, what would happen in your State if you had to move toward licensure of a dry storage facility or a storage facility to handle beyond your current capacity?

Mr. SHERMAN. Senator, that is a good question, and I think that you are all aware of the contentious and disastrous experience that happened in Minnesota. They claim that that was the most contentious issue that they remember in their State legislature for years and years. It is possible that something like that can happen in Vermont. And we hope to avoid that, but we cannot say. So the answer is it has the potential to be extremely contentious.

Senator JOHNSTON. That is where the price of building dry cask storage was building more windmills?

[Laughter.]

Mr. SHERMAN. Yes, that is correct.

Senator CRAIG. Mr. Sherman, how important is nuclear energy to the Northeast?

Mr. SHERMAN. As you know, Senator, nuclear energy is very important to the Northeast. Vermont gets about one-third of its energy from nuclear power, and the Vermont Yankee Nuclear Powerplant produces 85 percent of the electricity generated in Vermont.

Senator CRAIG. Well, those are the facts. We are faced with realities. I think the committee has done an excellent job this morning, and all of you have in the testimony you have delivered. And especially to those of you who are recognized as experts by your experience and your involvement in the transportation of spent nuclear fuel and nuclear waste, Bennett Johnston is correct, you are correct, to our knowledge no cask has ever been violated or damaged in a way that would cause leakage, and the record is absolutely safe. And somehow we all have to effectively communicate that.

It is interesting that in Nevada, while my legislation chooses the expensive route to avoid the politics of Las Vegas, most all activities that have occurred at the Nevada test site that have been transported by rail that were designed to go boom went right through Las Vegas and over Hoover Dam. And in those days the politics allowed that an nothing was said and nothing happened. Of course nothing happened. It is important that I say that for the record. This is the politics of it and not the reality of it or the science of it or the engineering of it or the safety of it, by history and by recorded fact.

Gentlemen, thank you very much for being with us today, and your patience has been appreciated. We do appreciate your testimony and look forward to working with you to resolve this problem.

The committee is adjourned.

[Whereupon, at 12:45 p.m., the hearing was adjourned.]

APPENDIX

ADDITIONAL MATERIAL SUBMITTED FOR THE RECORD

U.S. NUCLEAR REGULATORY COMMISSION,
Washington, DC, December 20, 1995.

Hon. FRANK H. MURKOWSKI,
Chairman, Committee on Energy and Natural Resources, U.S. Senate, Washington, DC.

DEAR SENATOR MURKOWSKI: I am pleased to provide written testimony for the record for the December 14, 1995 hearing before the Senate Energy and Natural Resources Committee on S. 1271, the "Nuclear Waste Policy Act of 1995." Should you have any questions, please contact me or Dennis Rathbun, Director, Office of Congressional Affairs.

Sincerely,

SHIRLEY ANN JACKSON,
Chairman.

[Enclosure.]

STATEMENT OF SHIRLEY ANN JACKSON, CHAIRMAN, U.S. NUCLEAR REGULATORY COMMISSION

Mr. Chairman, members of the Subcommittee, I am pleased to accept your invitation to provide written testimony for the record, on behalf of the Nuclear Regulatory Commission, addressing S. 1271, the "Nuclear Waste Policy Act of 1995," pending before this Committee. The Commission agrees with the fundamental approach of S. 1271, although some aspects are of concern to us.

We believe that S. 1271 contains the fundamental elements of an integrated high-level waste management plan needed for the protection of public health and safety and the environment. They are interim on-site storage, centralized interim off-site storage, and deep geologic disposal, together with a transportation mechanism to tie the elements together. In our view, S. 1271 recognizes that the overall, long-term success of this nation's program to manage spent fuel and other high-level radioactive waste is dependent on finding a permanent solution to the safe disposition of this material.

With respect to interim on-site storage, we note that NRC has approved seven dry cask systems for storage of spent fuel at commercial nuclear power reactor sites using a general license process authorized in Section 133 of the existing Nuclear Waste Policy Act, as amended (NWPA). We are therefore pleased that a provision for generic certification of storage systems is included in S. 1271 so that the Commission can continue to build on its existing program.

We believe the provision in S. 1271 for centralized interim storage of spent fuel has several positive features. Although indefinite storage in an above-ground facility is inconsistent with the Commission's Waste Confidence decision which includes a mined geologic repository, continued at-reactor storage for an interim period would protect public health and safety. However, a centralized interim storage facility, when compared to dispersed storage at more than 74 sites throughout the country, would allow for a more focused inspection and surveillance program by both DOE and NRC and would also offer operational and programmatic benefits on DOE's program for accepting waste from utilities. Therefore, because there are clear advantages to centralized interim off-site storage, we believe that establishment of such a facility under S. 1271 could partially resolve one of the most vexing issues facing this nation. However, the NRC takes no position as to where a centralized facility should be located. Thus, we support and recommend Congress include the provision in S. 1271 on Waste Confidence which would base that confidence not only on DOE's obligation to construct and operate an interim storage facility, but also on its obligation to develop the integrated spent fuel management system.

The Commission does not object to the S. 1271 provision for issuance of a license for centralized storage in phases, to accommodate expansion of storage capacity. However, it should be recognized that some generic licensing considerations and siting factors are essentially independent of storage capacity and will need to be met by DOE for all licensing phases.

Nonetheless, we would like to comment on some aspects of S. 1271 related to centralized interim storage that would impact our normal regulatory process. In particular, our concern relates to the constraints that the proposed timeline in S. 1271 might place on our normal procedures for public involvement in our regulatory process. Public participation is a fundamental part of NRC rulemaking and licensing actions and key to retaining public confidence in our process. S. 1271 would require the Commission to license the phase one centralized interim off-site storage facility within 16 months after submittal of an application. S. 1271 would also require the Commission to issue regulations within 18 months of the Act setting criteria for storage at the centralized facility of foreign research reactor and naval reactor spent fuel as well as high-level wastes from defense activities. If the public were to exercise full participation rights under our procedures, it is doubtful that we could accomplish these activities in the time allowed.

The Commission strongly supports including in S. 1271 deep geologic disposal as an essential element of an integrated program. The Commission continues to believe that deep geologic disposal is a sound and technically feasible solution to the problem of final disposition of spent nuclear fuel and other high-level radioactive wastes. Further, we are confident that the Commission will be able to determine with reasonable assurance that spent fuel and other high-level wastes can be disposed of safely in a geologic repository assuming that we get a high quality license application and that NRC can maintain its technical capabilities for licensing deep geologic disposal in the current budgetary environment. In addition, we agree with provisions in S. 1271 which revoke the DOE siting guidelines. We believe this will allow DOE to focus its repository efforts on developing the high-quality license application that we will need.

Further, we support provisions in S. 1271 on the scope of NRC's National Environmental Policy Act (NEPA) responsibilities for disposal that (consistent with existing law) direct NRC to adopt DOE's EIS, to the extent practicable, in the repository licensing proceeding. (The Commission also supports the bill on specifying the scope of NRC's EIS, requiring the generic consideration of transportation impacts, and identifying the issues that should not be considered by the Commission under NEPA for interim storage).

With respect to the standard in S. 1271, the Commission does not object to a single, overall performance standard for the repository for a 10,000-year period following commencement of repository operations. The Commission notes the standard in S. 1271 of an annual effective dose of 100 mrem or 1 mSv, and views that standard as consistent with the protection of the public health and safety. Within this limit, to enhance confidence that the standard will be met, the Commission would expect DOE to evaluate alternatives among the major repository design features. Further, in order to facilitate implementation of the standard specified in S. 1271, some clarification in the legislation may be necessary to allow the Commission's use of subsystem performance measures to evaluate major repository system features from the standpoint of whether they contribute to or detract from overall system performance.

Further, as to section 206(b), which directs the Commission to amend its regulations on disposal "upon the completion" of the phase one storage facility licensing, we believe the Commission should have the flexibility to amend its regulations at any time in order to assure needed regulatory stability for development of DOE's repository program.

The Commission believes that progress on the repository program is very important to the overall success of the nation's high-level waste program which, as previously noted, depends on finding a permanent solution to the safe disposal of these wastes. In this regard, the Commission believes that S. 1271, which gives budget priority to storage, should recognize that DOE must control storage costs in a cost-efficient manner in order to assure adequate funding for repository development.

Finally, the Commission has a comment on the provision in S. 1271 concerning compliance with other laws. Specifically, under that provision, all actions on the integrated program are to be governed by the Atomic Energy Act, the Energy Reorganization Act, the Hazardous Materials Transportation Act and the Nuclear Waste Policy Act. The provision does not mention NEPA. However, as previously noted, S. 1271 also includes other specific provisions on Commission responsibility under NEPA. Therefore, it may be useful to clarify whether Congress expects the Commission to use its NEPA responsibility under S. 1271 to require compliance with environmental and other statutes such as the Endangered Species Act or the Clean Water Act.

In summary, the Commission agrees that S. 1271 constitutes an appropriate program for the permanent disposition of high-level radioactive waste by providing an integrated spent fuel management system that contains the necessary fundamental elements—on-site interim storage, centralized off-site storage, and deep geologic disposal. There is a need for both statutory and institutional stability for our nation's high-level waste program so that the program can proceed in an orderly and business-like fashion. The Commission believes that, when coupled with funding sufficient to maintain progress in all phases, S. 1271 can and will provide the necessary stability for the nation's high-level radioactive waste program. We appreciate the opportunity to provide our views.

THE COMMONWEALTH OF MASSACHUSETTS,
EXECUTIVE DEPARTMENT,
Boston, MA, December 12, 1995.

Hon. FRANK H. MURKOWSKI,
Chairman, Committee on Energy and Natural Resources, U.S. Senate, Washington, DC.

DEAR SENATOR MURKOWSKI: I am writing in support of the Nuclear Waste Policy Act of 1995 (S. 1271), which was introduced by Senator Larry Craig. As you well know, S. 1271 is the companion bill to H.R. 1020, which has received broad bipartisan support in the House of Representatives.

From an environmental standpoint, storing nuclear waste at one centralized facility is certainly preferable to keeping waste at over 100 different plants across the country. In addition, removing waste in 1998 rather than 2010, as currently projected by the Department of Energy (DOE), will allow the owners of the two sites in Massachusetts (in Plymouth and Rowe) to return them to green fields much sooner.

S. 1271 addresses many of the concerns shared by the citizens of Massachusetts with regard to nuclear waste management. The bill clarifies and reaffirms the Department of Energy's obligation to accept commercial spent nuclear fuel starting in 1998, and specifically calls for the construction of both an interim storage facility and a permanent repository on federal lands in Nevada. S. 1271 also calls for the transfer of responsibility for spent fuel at plant sites from utilities to the DOE, and ensures the transportation, infrastructure and planning are in place to ship fuel safely to the interim and permanent facilities.

To date, ratepayers nationwide have contributed nearly $11 billion to the Nuclear Waste Fund for the construction of both interim and permanent facilities, yet the DOE has not come close to completing either project. In fact, the DOE has essentially abdicated its responsibility by refusing to acknowledge its legal obligation to begin accepting spent fuel by 1998. Instead, the Department claims an obligation to accept waste only upon the completion of the permanent repository, which has no predetermined deadline. To compound matters, less than half of the $11 billion has actually been appropriated, while the balance has been used for other unrelated purposes. Electricity customers in the Commonwealth of Massachusetts have paid over $300 million, and electricity customers across New England have paid over $1 billion. These customers deserve some return on their huge investment. Thus, it is essential that S. 1271 be enacted.

H.R. 1020 is expected to be considered by the full House before the end of 1995. It is very important to achieve a similar consensus for S. 1271 in the Senate. I respectfully urge you to support this legislation and to help persuade your colleagues to join in this vital effort.

Sincerely,

WILLIAM F. WELD,
Governor.

STATEMENT OF SCOTT NORTHARD, PROJECT MANAGER, MESCALERO/UTILITY FUEL STORAGE INITIATIVE, ALBUQUERQUE, NM

INTRODUCTION

Mr. Chairman and Members of the Senate Energy and Natural Resources Committee, thank you for convening a hearing on S. 1271, the Nuclear Waste Policy Act of 1995, introduced by Senator Larry Craig on September 25, 1995. The Mescalero/Utility Fuel Storage Initiative (the Mescalero/Utility FSI) is submitting this statement for the hearing record because we believe that maintaining the option of a privately-financed temporary spent-fuel storage facility is critical to the development of a credible and comprehensive nuclear waste policy. Moreover, we believe that the Mescalero/Utility FSI represents the most advanced policy and planning for a temporary facility in the private or public sector anywhere in the world.

The Mescalero/Utility FSI combines the efforts of twenty-one electric utilities and the Mescalero Apache Tribe of New Mexico to construct a temporary spent-fuel storage facility on the Mescalero Tribal lands in New Mexico. The facility is intended to store spent nuclear fuel; i.e., the small ceramic nuclear pellets that must be removed from a nuclear reactor on a periodic basis after their usefulness is exhausted. Currently, spent nuclear fuel is stored primarily in water-filled pools at nuclear power plant sites, although some storage occurs in above-ground, non-water storage technologies licensed by the Nuclear Regulatory Commission.

The private spent-fuel option is important for a number of reasons. First, the Mescalero/Utility FSI will expand significantly the nation's temporary storage capacity at a centralized location. Second, the Mescalero/Utility FSI will serve as an essential contingency for some utilities to account for continuing delays in the Department of Energy (the DOE) programs. Third, the Mescalero/Utility FSI provides a true market-based check against and standard by which a DOE temporary facility can be measured and evaluated. For the Mescalero/Utility FSI to succeed, it must be cost-effective, comply with all safety and environmental regulations and stay on its completion date time frame. The best guarantee that each of these objectives can be achieved is the investment by private investors in the Mescalero/Utility FSI.

Mr. Chairman, as you and the Members of the Energy Committee know, these objectives are often difficult to achieve in DOE projects. We believe that the Mescalero/Utility FSI is essential for providing an "honest" benchmark for a DOE project. Moreover, the Mescalero/Utility FSI will proceed at the same pace or even ahead of the DOE project, and many of the issues that will face the DOE project will be resolved first at Mescalero such as technology development, public education and others. To an extent, the Mescalero/Utility FSI already is serving this important role. In fact, the DOE reduced the cost estimate for it interim storage facility by almost fifty percent—from $1.1 billion to $567 million—after evaluating the cost estimates for the Mescalero/Utility FSI project. Unfortunately, the DOE cost estimate remains over twice as high as the Mescalero/Utility FSI project at this stage.

The testimony that the Committee heard at the hearing demonstrated the need for a temporary facility for storage. The Mescalero/Utility FSI is an important, and we would argue, essential alternative. The facility will be used only as a storage site; no handling or processing of individual fuel assemblies will take place. The utilities that transport spent fuel to the site will retain ownership of the fuel. Moreover, Mescalero/Utility FSI will use only storage and transportation technologies that have been approved by the Nuclear Regulatory Commission (the NRC) and are compatible with the DOE permanent repository. The Mescalero/Utility FSI also will utilize design, operation and decommissioning standards that are similar to existing utility storage efforts in order to minimize costs and maximize reliability.

FACILITY DESCRIPTION

When complete, the overall facility will resemble an industrial park. Several large buildings plus transportation facilities will be constructed. The facility will include three primary buildings, in addition to the storage area. The buildings will be designed to reflect architecturally the area's natural setting with a minimum of environmental intrusion. Occupying the largest land area on the site, the fuel storage area will consist of a series of concrete and steel compartments which will each hold a steel canister containing spent fuel. Current canister designs allow for the storage of 24-52 spent fuel assemblies, depending on the type of assembly. There will be approximately 2,000 compartments plus provisions to expand the facility if additional storage is needed. The fuel storage-area will be closely monitored with security fencing and lighting, intrusion-detection devices, closed-circuit television and radiation-monitoring systems.

The operations and maintenance building will be the central area of activity at the facility. An arriving shipping container with a loaded fuel canister will be positioned in the building using an on-site locomotive. A canister-transfer system will remove the canister from the shipping container and transfer it to a storage compartment.

The security and health-physics building will house the staff and equipment for general security operations, radiation- and environmental-monitoring operations, and fire and emergency medical services.

The administration building and attached visitors center will house up to 75 full-time staff and professional personnel. It will include space and equipment for site management and administrative activities, licensing, engineering, quality assurance, records management, communication, emergency response and personnel training.

The cost to construct and equip the facility is estimated at $99 million. Cost for design, engineering, and license of the facility is estimated at $35 million.

SPENT-FUEL TRANSPORTATION TO THE MESCALERO SITE

Canister loading and all handling of spent fuel assemblies will be done at the power plant prior to transporting it to the Mescalero facility. The spent fuel will be transferred from utility storage into NRC-approved storage canisters and sealed inside.

Each canister will then be placed in a large shipping container and transferred by rail to the Mescalero facility. Upon arrival, the canister will be removed from the shipping container and placed in a shielded-fuel storage compartment.

SITE DESCRIPTION

After study of several potential locations, the Tribal Council selected the Three Rivers Area for further facility-siting consideration. The Three Rivers Area is in the desert lands located in the northwest corner of the Reservation. It is approximately 20 miles from the nearest town and within 10 miles of a well-maintained rail line operated by Southern Pacific Railroad. The overall site will encompass approximately 250 acres of the 460,000-area Mescalero Apache Reservation. Most of the site area will be used as a restricted-access buffer zone with the actual fuel-storage facility occupying only about 35 acres or 14 percent of the total site. The DOE has already conducted a preliminary site-suitability assessment for this area and determined that is was acceptable for further evaluation.

REGULATIONS

The facility design will meet all applicable federal regulations, as well as all tribal codes and laws. The facility will receive a site-specific license from the NRC. Obtaining a license requires review and approval of the site and the facility design—a rigorous review process similar to that required to license a nuclear power plant. Factors to be considered include site geology, facility design, environmental assessment, quality assurance, technical and financial qualifications, emergency response planning, personnel training, radiological protection, and physical security. An environmental impact statement will be developed as part of the process. The estimated review time for the NRC application is 24 to 36 months. The term of the NRC license will be 20 years with provisions for an extension to a total period of up to 40 years. Regular on-site environmental monitoring will be conducted by highly trained staff analysts, with oversight by the NRC. Air, water and land samples will be taken frequently to ensure the environmental integrity of the facility and surrounding lands.

OPERATIONS

Canister handling and the number of shipments will be minimized to keep storage costs as low as possible. Large capacity canisters and shipping containers will be used to minimize costs, number of workers required, number of shipments, and to limit occupational radiation exposure. The site will be designed to accept approximately 100 shipments each year. Following the temporary storage period (estimated at 20 years) the canisters will be retrieved from the storage compartments in the same manner they were transferred initially. The canisters will be placed in shipping containers for transfer to DOE's geologic repository for permanent disposal.

CONCLUSION

Mr. Chairman, for many reasons progress on a DOE facility has been painfully slow. Under the best circumstances, the completion of a facility is both tens of millions of taxpayer dollars and many years away. The Mescalero/Utility Fuel Storage Initiative offers a temporary solution for many utilities to a pressing national concern—one which directly affects twenty percent of our nation's electric energy needs. Using free market principles and the path of private enterprise, we can solve this problem safely, effectively and responsibly.

The Mescalero/Utility FSI project will meet or exceed all licensing and regulatory requirements established by the Nuclear Regulatory Commission and a host of other federal agencies. In the end, it will save dollars for ratepayers as well as taxpayers, primarily because it creates the economical advantage of a centrally located storage facility, encourages cost effective nuclear plants to continue operating, and allows permanently shut-down plants to be decommissioned—without having to serve as storage sites. When completed, it will be a shining example of how government can find practical, responsible answers to our nation's energy needs, and workable ideas for our nation's future when it relies on the innovation of the private sector. For all these reasons, the Mescalero/Utility FSI is an option that must be preserved.

STATEMENT OF THE PRAIRIE ISLAND INDIAN COMMUNITY

A. INTRODUCTION

Chairman Murkowski and members of the Senate Energy and Natural Resources Committee, thank you very much for allowing the Prairie Island Indian Community this opportunity to submit written testimony to you and your colleagues.

The Prairie Island Indian Community has been proactive on nuclear waste issues. The Prairie Island Indian Community's interest and activism in nuclear waste issues arises from establishment of a nuclear power generating plant adjacent to the Prairie Island Indian Community since 1973: To date, we do not know of anyone in the United States who lives so close to a nuclear facility.

Currently, our neighbor, the nuclear power plant, has begun to store spent nuclear fuel in above ground casks only three blocks from the homes, school and business of our community. Our tribal council has been monitoring efforts to develop a national solution for nuclear waste storage.

We want to voice our concerns about the bill before the Senate Energy and Natural Resources Committee, S. 1271, Amendments to the Nuclear Waste Policy Act. We ask that this legislation be amended to allow advance consultation with Indian tribes throughout the transportation process, the siting process and wherever Indian tribes are impacted either directly or indirectly. Further, we ask that Indian tribes be included wherever local units of government and states are mentioned in the legislation with regard to technical assistance and funds.

Additionally, we desire an amendment to ensure that once spent nuclear fuel is transported and stored at interim/permanent sites that at no time will the spent fuel be sent back to its place of origin. Also, we ask that Congress continue to fund the multi-purpose canister (MPC) system this session. Finally, we ask this Committee and Congress as a whole to continue every effort to establish both an interim and permanent storage site for the health and safety of all the citizens of this Country.

We ask these things as a matter of the United States Government's "trust responsibility" toward American Indians, which is the unique legal and moral duty of the United States to assist Indians in the protections of their property and rights. This trust responsibility arises out of treaties, statutes, executive orders, legal precedence, the United States Constitution and the course of dealings between the United States Government and Indian tribes. The inclusion of Indian tribes in the processes of this legislation is a fulfillment of this responsibility.

B. HISTORY OF THE PRAIRIE ISLAND INDIAN COMMUNITY

The Prairie Island Indian Community is a federally recognized Indian tribe organized under 25 U.S.C. §476 and is governed under the terms of the Constitution and By-Laws adopted by tribal members on May 23, 1936 and approved by the Secretary of the Interior on June 20, 1936. Its members are Mdewakanton and Wahpekute descent of the Dakota Sioux. There are approximately 500 enrolled members; one hundred and sixty (160) members live on the reservation. The Prairie Island Indian Community is located on Prairie Island, which is formed at the confluence of Vermillion and Mississippi rivers in southeastern Minnesota, approximately 40 miles southeast of Minneapolis. The reservation is on low-lying Mississippi River bottom land surrounded by bluffs. The reservation is approximately 395 acres. In 1992, the United States Army Corps of Engineers granted the tribe exclusive use of an additional 225 acres of Corps land at the north end of Prairie Island to be used as if it were Prairie Island Indian Community's land.

C. HISTORY OF NUCLEAR POWER ON PRAIRIE ISLAND

Northern States Power (NSP) has operated a nuclear power plant located three city blocks from the Prairie Island Indian Community since 1973. The original purchase of land for the power plant was through non-tribal members residing adjacent to the reservation. The tribal government was not included in any aspect of the purchase or negotiations. Since the plant went on-line, Prairie Island Indian Community members have been concerned about nuclear power's impacts on the environment and health of the people. The following events have heightened the Prairie Island Indian Community's concerns: two radiological leaks, storage of nuclear waste inside the plant and outside the plant in dry casks, and the transportation of hazardous materials to and from the plant on railroad lines adjacent to the reservation. The lack of full tribal participation in emergency evacuation planning and preparedness has augmented fears about living next to a nuclear power plant.

The tribe receives no benefit from the nuclear power plant. The Prairie Island Indian Community receives no revenue or compensation from NSP. The Prairie Island Indian Community does not even receive electricity from this power plant. The City of Red Wing (located approximately 10 miles southeast of Prairie Island), however, receives approximately $28,000,000 (Twenty- Eight Million Dollars) annually from NSP's property taxes alone. In June 1971, Red Wing annexed Prairie Island itself to collect the property taxes.

In 1994, NSP was successful in obtaining legislative approval to store spent nuclear fuel in dry casks. NSP was given permission to use a total of 17 casks (each cask holds 40 spent fuel assemblies). As NSP was quickly running out of pool storage space, the additional storage capacity offered by the dry casks was to keep the plant operational. The long and divisive legislative session polarized the Tribe and the surrounding communities. Most surrounding communities supported the legislation because they believed that it was the only way to keep the plant open and preserve jobs. Although the Prairie Island Indian Community recognizes the importance of preserving local jobs, our concerns centered around the health and safety effects the casks would have on the human beings who live here.

Minnesota is the only state that has required a utility to seek legislative approval to store nuclear waste in dry casks. Because the casks are stored on-site, within the plant boundaries, NSP first had to receive a Certificate of Need from the Minnesota Public Utilities Commission. The case was subsequently referred to the Office of Administrative Hearings. While the Administrative Law Judge reviewing the case stated that the storage of waste in dry casks appeared to be a "reasonable, safe, and cost effective means of storing high-level radioactive waste on a basis," he also stated that the likelihood that the dry cask storage would become permanent is so great as to require legislative authorization.

All of these activities have placed a strain on tribal relations with state and local governments. The tribe has just now begun to work together with the other governments on common issues and concerns. The first dry cask was placed outside, on a concrete storage pad, on May 15, 1995. The second dry cask was placed outside on November 13, 1995. The third cask was placed outside on December 21, 1995. As more and more casks are filled and placed on the storage pad, the expectation that the existing on-site storage facility will become a permanent facility increases and expectations for a national solution diminish.

D. TRUST RESPONSIBILITY

The United States has a trust relationship with the Prairie Island Indian Community. The United States' trust responsibilities include protecting tribal interest and vindicating tribal rights. The United States has a responsibility to protect the environment, human health, cultural resources, and property interests in Indian Country.

The United States has a direct obligation to assure us that the waste currently stored at the plant poses no risk to our health and safety and our economic security. The Nuclear Regulatory Commission (NRC) licenses the plant and has oversight for all plant activities. The Department of Energy (DOE), as mandated in the Nuclear Waste Policy Act, is obligated to begin accepting spent nuclear fuel by January 31, 1998. By fulfilling this obligation, the DOE will also be fulfilling its trust responsibility to the Prairie Island Indian Community.

Further, this trust responsibility has been acknowledged in a recent Memorandum of Understanding (MOU) among the Bureau of Indian Affairs, the Environmental Protection Agency, Federal Emergency Management Agency, Indian Health Service, National Park Service-Archaeological Assistance Division, Nuclear Regulatory Commission, U.S. Fish and Wildlife Service and other federal entities and the Prairie Island Indian Community. Though not all the agencies have signed the MOU to date, there have been three successful meetings of agency representatives regarding their responsibilities under the MOU.

Additionally, several United States Presidents have acknowledged the United States' trust responsibility, by supporting the policy of government-to-government relationships with Indian tribes. Nuclear waste policy is no exception. It is absolutely necessary that tribes are consulted on nuclear waste policy. Not only are state and local governments affected by nuclear waste but tribal governments are as well.

The Prairie Island Indian Community is affected whether the nuclear waste stays or goes. The tribe is also concerned about transportation of nuclear waste near the reservation. The Chicago Milwaukee St. Paul Pacific Railroad runs along the reservation boundaries. The proposed legislation does not address points of origin nor address tribal consultation to tribes affected along the route. We, and other Indian tribes, must be consulted in either event.

E. CONCERNS ABOUT LONG-TERM STORAGE OF NUCLEAR WASTE

Prairie Island Indian Community supports the expeditious and safe removal of spent nuclear fuel from Prairie Island. We do not agree with the Congressional Finding in the bill that "spent nuclear fuel can be safely stored at reactor sites." We also support a national solution to the issues of storage of spent nuclear fuel.

Health and safety

The Prairie Island Indian Community did not create the nuclear waste storage problem but we are forced to live next to it. We do not benefit from the production and storage of spent nuclear fuel, yet we bear the greatest risk. Since the plant went on-line, we feel that long-term exposure to low-level radiation has caused cancer and other health effects on our people. No federal agency has ever conducted a health assessment or health study on Prairie Island, yet we are constantly told that the power plant and dry cask storage are safe and pose no threat to our health and safety. The MOU includes health assessments and studies but, to date, these tasks have not been completed nor has the assistance been provided regarding these tasks much to the frustration of the Prairie Island Indian Community.

Prairie Island Indian Community members face additional health concerns with regard to the storage of spent nuclear fuel in dry casks at the plant. The storage of spent nuclear fuel in dry casks emits additional amounts of radiation in the form of neutrons and gamma rays into the environment. If no federal solution to the problem of nuclear waste storage is found, the human exposure to such radiation is increased. Without a federal solution to this problem, the so-called temporary dry cask storage facility at Prairie Island will become a de facto permanent storage facility.

Our people do not want to reside next to a permanent storage facility. Because of the threat of catastrophe at the nuclear facility on Prairie Island, many of people wish to relocate to a safe and secure environment. The Prairie Island Tribal Council is pursuing all viable options for relocation. That means obtaining lands to relocate those members who want to move and build a community free of nuclear waste.

We believe that our children are an investment in the future. We are securing our heritage through the establishment of the Prairie Island Indian Charter School. It is an independent public school for kids in preschool through grade 12. The school currently has 50 students. Because the school is within five blocks of the nuclear power plant and dry cask storage, there is concern that the children will suffer from the effects of long-term exposure to radiation.

The federal government has a trust responsibility to protect the health and welfare of Indian tribes. Protecting our people from harmful impacts demands that the federal government expeditiously and safely remove nuclear waste from Prairie Island.

The land that the Prairie Island Indian Community and the nuclear power plant share is not very big. For economic reasons, the Prairie Island Indian Community has utilized existing buildings for the school and continues the process to purchase other land farther from the nuclear power plant.

Economic development

In addition to the health and safety risks, the nuclear power plant threatens the financial security of the tribal business, Treasure Island Casino and Bingo. Not only are our people negatively impacted by the existence of the power plant and storage casks, but our customers of the tribal enterprise are impacted as well.

We, as tribal leaders of a sovereign Indian nation, are responsible for ensuring the health and well-being of our people and tribal resources. The tribe also has a responsibility to protect its 1500 employees and its many guests who visit the reservation every day.

The threat of nuclear power generation and nuclear waste storage has a chilling effect on the economic growth of this Prairie Island Indian Community. The limited land base also impacts future economic growth on our reservation. Current tribal enterprises include the Treasure Island Casino and Bingo, recreational vehicle park and marina on the Mississippi River. Potential business partners, investors and financial parties are cautious about working with the Prairie Island Indian Community because of the perceived risks from the nuclear power plant and storage site.

Transportation concerns

The Prairie Island Indian Community is also concerned about the transportation of nuclear waste from the plant to either an interim or permanent facility. Since spent nuclear fuel will be transported on the railroad running adjacent to the reservation, the Prairie Island Indian Community must be informed of and fully involved with all waste shipments from the plant. Full involvement includes: advance notice of shipments, technical and financial assistance for emergency preparedness activities on the reservation, and an opportunity to technical provide input on shipping campaigns.

We recognize that the federal government has an obligation to the public utilities to develop a solution to this national problem. It is imperative that the federal government arrive at solutions that also addresses issues important to Indian tribes. As the Prairie Island Indian Community will be impacted by shipments of spent nuclear fuel from NSP's plant, whether to an interim or permanent facility, we expect to be fully involved and consulted well in advance of any shipments.

F. RECOMMENDATIONS

1. Amend S.F. 1271 to require advance consultation between the Secretary of Energy and her designees with Indian tribes throughout the transportation process, the siting process and wherever Indian tribes are impacted, directly or indirectly.
2. Amend S.F. 1271 wherever local units of government and states are mentioned, to also include Indian tribes in technical assistance and funding provisions.
3. Amend S.F. 1271 to prohibit spent nuclear fuel which has been removed to an interim/permanent storage site from being returned to its place of origin. We would like assurances that once nuclear waste is removed from Prairie Island to an interim storage facility that it will not be returned. The bill provides for storage at an interim facility for 100 years yet does not contain assurances that waste will not be sent back to its point of origin if a permanent repository has not been licensed.
4. Congress continue to fund the MPC System. We understand that funding for MPC System may be in jeopardy. The MPC system was to provide a safe mechanism for both storing and transporting spent nuclear fuel to either a interim or permanent facility. If congress intends for the DOE to meet the January 31, 1998 deadline, the funding to fully develop this technology must be restored.
5. Congress be diligent in establishing an interim/permanent storage site.
6. Congress respect and live up to its trust responsibilities. For the United States to fulfill its trust responsibility, it must work with tribes on a government-to-government basis to ensure full tribal participation. The Department of Energy must provide us with technical and financial assistance so that we are adequately prepared for all shipments of spent nuclear fuel traveling adjacent to the Prairie Island Indian Community and any impacted Indian Nation.

Further, in upholding its trust responsibility, we ask Congress to negotiate with affected Indian tribes regarding land, compensation, and health and safety issues.

7. Amend S.F. 1271 to include in the definition of an affected Indian tribe, any Indian tribe—"whose trust land abuts any nuclear power generating plant and any above ground storage facility."

ALPER & MANN, P.C.,
Washington, DC, December 6, 1995.

Hon. FRANK H. MURKOWSKI, *Chairman,*
Hon. J. BENNETT JOHNSTON, *Democratic Chairman,*
Senate Energy and Natural Resources Committee, Senate Dirksen Office Building, Washington, DC.

Re: Pending Legislation Governing the Transportation of Nuclear Fuel and Radioactive Waste

DEAR CHAIRMEN: This letter is being written on behalf of the Railway Labor Executives Association and the United Transportation Union. The constituent members of the RLEA are attached hereto. The RLEA members and the UTU represent nearly all of the railroad workers who will be potentially exposed to radiation which might result from a train accident. These include the operating crafts, certain shop crafts, and the maintenance of way employes.

The risks are simply too great for Congress not to specifically address the safety issues inherent in the movement by rail of spent nuclear fuel and radioactive waste. It is estimated that 85-90% of such movements will be by rail transportation. Officials in the Government and private industry have in effect said "trust me, the shipments will be safe." However, no one in this country knows better than the railroad workers who operate the trains, inspect and maintain the equipment and track, that railroad transportation, while generally safe, raise numerous safety concerns. One need only to look at the sabotage caused train derailment in Arizona in October to realize the vulnerability of train shipments.

We believe that the legislation currently being considered by Congress for the rail shipment of spent nuclear fuel and radioactive waste is deficient in some respects. For example, there are no specific provisions requiring the training of all crews connected with such movements, no increased inspections and testing of the train and tracks, train make-up is not addressed, no speed restrictions, routing is not restricted to upgraded track, security needs should be spelled out so that everyone involved in the movements will know how to deal with emergencies, and placarding should be improved.

We recognize that regulations exist which cover some of the above areas. However, they are scattered over several thousand pages of regulations, and therefore it is extremely difficult for rail workers to know the requirements. All of the regulations covering nuclear fuel and radioactive materials being shipped by rail should be located under one separate part of the Code of Federal Regulations, even though some of the provisions may be administered by different agencies.

We understand that your Committee's hearings on December 14, will not focus specifically on the safety issues related to railroad transportation, and that you will at a later date consider those mailers. At such time we would appreciate the opportunity to present oral testimony and provide your Committee with an in depth analysis of the issues raised in this letter.

Sincerely,

LAWRENCE M. MANN,
Attorney for RLEA and UTU.

INTERNATIONAL BROTHERHOOD OF TEAMSTERS,
Washington, DC, December 11, 1995.

Hon. RICHARD GEPHARDT,
Minority Leader, U.S. House of Representatives, Washington, DC.

DEAR REPRESENTATIVE GEPHARDT: The International Brotherhood of Teamsters, on behalf of its 1.5 million members, urge you to support H.R. 1020—the Nuclear Waste Policy Act of 1995, which would provide for the centralized management and disposal of spent fuel from commercial nuclear power plants. This legislation is good policy to safeguard not only the environment but the communities which surround our nuclear power plants. The expeditious passage of this bill will help fulfill the promise that the federal government made with the nation's electricity customers to build safe nuclear waste storage facilities.

With the passage of the Nuclear Waste Policy Act of 1982, a fund was created that generates $600 million annually from electric utility customers to federally manage nuclear waste. Little has been accomplished since the Act was passed. To this day the Department of Energy (DOE) is not able to meet its obligation to begin accepting and storing spent fuel from nuclear power plants.

The solution to (his problem lies within the provisions of H.R. 1020, which was reported out by the Commerce Committee by a 30-4 vote, clearly demonstrating bipartisan support. The legislation would provide the DOE with the tools it needs to fulfill its obligations by 1998. The bill addresses premature shutdowns that would adversely affect the communities which house the nuclear facilities and the burdensome costs of on-site storage of the spent fuel which would be unfairly cost shifted to the electric customers of these plants.

The funding of the nuclear waste program is shouldered by the user who pays a flat 1/10 cent per kilowatt hour fee on nuclear generated electricity. Since 1982, the Nuclear Waste Fund has collected $11 billion and spent only $5 billion. At least $200 million collected per year is used to mask the deficit. The Commerce Committee bill would address this inequity by assessing the waste fund payments based on the amount appropriated by Congress. This would reinforce the federal government's commitment in resolving this issue through honest fiscal policy.

The passage of this legislation is vital to the IBT and its membership who perform nuclear waste cleanup and represents a long overdue commitment not only to the safety of the workers involved but also to the American people.

Sincerely,

RON CAREY,
General President.

STATEMENT OF DR. ANDREW C. KADAK, PRESIDENT AND CEO OF YANKEE ATOMIC ELECTRIC COMPANY

INTRODUCTION

Yankee Atomic Electric Company submits this testimony in support of S. 1271, "The Nuclear Waste Policy Act of 1995," and urges that it be passed this session of Congress. S. 1271 is an important bill for the Commonwealth of Massachusetts and the New England region.

Nuclear power has been a significant source of safe and reliable electricity for the New England region since the Yankee Nuclear Power Station in Rowe, Massachusetts began operating in 1960—35 years ago. Currently, the region's eight operating nuclear power plants provide on the order of 40% of the electricity generated in New England—the highest percentage of any region in the U.S. This comparatively high reliance on nuclear energy (the national average is 20%) is projected by the New England Power Pool to continue for the foreseeable future.

This legislation is necessary to ensure that the Department of Energy (DOE) provides a timely solution to the high-level nuclear waste storage and disposal situation currently impacting the electric consumers of the state and region.

BACKGROUND

The U.S. Congress passed the Nuclear Waste Policy Act of 1982 (NWPA) in order to provide for the management and disposal of spent nuclear fuel and high-level radioactive waste. The Act, including amendments, directed the DOE to develop a nuclear waste management and disposal program and to begin taking spent nuclear fuel from commercial nuclear power Sites in 1998.

To pay for the program, Congress established a one-mill per kilowatt-hour fee on the electricity generated by nuclear reactors. Since passage of the NWPA, the customers of the electric utilities of Massachusetts and New England have paid over $1 billion into the Nuclear Waste Fund. As a nation, electric ratepayers have paid over $11 billion into the Fund.

Although the ratepayers have met their obligation, the federal government has not. The DOE's program to dispose of civilian nuclear waste is so far behind schedule that the current DOE estimate for having a permanent repository ready in 2010, is now in serious jeopardy. The DOE has no interim storage program or solution. If DOE does not begin taking spent fuel off New England nuclear power plant sites in 1998, as directed by the NWPA, several of the region's utilities will need to construct additional on-site spent fuel storage capacity, at ratepayer expense.

The failure of the federal government to remove the spent fuel and high-level waste on schedule and in accordance with its statutory and contractual obligations, will have an adverse economic impact on the ratepayers of Massachusetts and New England that is both unfair and unnecessary. This impact on the region is exacerbated by the situation at the shutdown Yankee Nuclear Power Station in Rowe, Massachusetts. Because projections of the DOE's published spent fuel acceptance rate schedule project that the last spent fuel assembly will not be removed from the Rowe site until 2018, electric ratepayers will have to pay additional and unnecessary storage costs on the order of $70 million. Obviously, some sort of expedited solution is needed for shutdown reactors like the Yankee Rowe plant.

Because the DOE's permanent repository will not be ready until at least 2010, a centralized federal interim waste storage facility needs to be established to add the storage problems confronting the nuclear power reactor sites in New England. Without this authorizing legislation, the DOE cannot expend funds to develop a centralized interim storage facility. S. 1271 proposes to develop and locate a federal interim storage facility at the U.S. government's Nevada Test Site by the end of 1998. Because this location was considered remote enough for the U.S. to explode over 300 nuclear devices between 1945 and 1974, this location is suitable to safely store spent nuclear fuel. A single interim federal site in Nevada is far preferable to multiple high-level nuclear waste sites in Massachusetts and other New England states.

SHUTDOWN REACTOR ISSUE

The need for priority acceptance for shutdown reactors has a clear historical basis. The April 1983 Federal Register Notice implementing the Standard Contract for Disposal of Spent Nuclear Fuel and/or High-Level Radioactive Waste (Standard Contract) clearly stated, "Article VIB.3 establishes two priorities for acceptance of SNF/HLW—one based on the age of the SNF/HLW and the second for the SNF/HLW from permanently shutdown rectors." The Federal Register Notice also stated, "Article VIB.3(b) gives priority to SNF/HLW from permanently shutdown reactors,

. . . This type of priority is necessary to prevent reactors from waiting 20 to 30 years to be decommissioned after they finish generating electricity."

The Yankee Rowe plant situation provides a clear example of the basis and need for such priority acceptance. Since Rowe's bent shutdown in 1992, Yankee has requested that the DOE grant priority acceptance of the 533 spent fuel assemblies (127 MTU) stored on-site and begin accepting the SNF and HLW in 1998 in accordance with the Standard Contract which provides for priority for shutdown reactors and specifically states, "Notwithstanding the age of the SNF and/or HLW, priority may be accorded any SNF and/or HLW removed from a civilian nuclear power reactor that has reached the end of its useful life or has been shut down permanently for whatever reason."

Yankees decommissioning plan for the Rowe plant provides that the site will be restored to a "green field" once the SNF/HLW is removed by DOE and the low-level radioactive waste is transported to an available disposal site. If DOE began to remove the SNF/HLW from Rowe beginning in 1998 on a priority basis, the site could be restored to a "green field" in a three year period.

However, the DOE has refused to grant Yankee priority acceptance. Even assuming DOE meets its contractual obligation to begin removing the waste in 1998, under the DOE's current acceptance rate schedule, called "OFF" (Oldest Fuel First), it is projected that DOE will not remove the last spent fuel assembly from the Rowe plant site until 2018, at the earliest. As a result, electric ratepayers will have to pay additional and unnecessary storage costs on the order of $70 million for Yankee to construct and operate a interim nuclear waste storage facility in Rowe, Massachusetts. This can be prevented if the DOE is directed by Congress to remove the waste from shutdown reactors sites such as the Rowe, Massachusetts site, on an expedited basis.

The situation at Yankee Rowe is also exacerbated by the fact that DOE's existing waste acceptance rate schedule shows the DOE coming to the Yankee Rowe site in each of the first four years of its "OFF" removal program, but only to remove fewer than a third of the spent fuel assemblies! With only 533 assemblies on the site which could be removed with 23 rail shipments, it is neither cost effective nor prudent for DOE not to remove all the waste within the first couple of years of the removal program. Priority acceptance was always intended as necessary to prevent permanently shutdown reactors like Yankee Rowe from having to wait decades to be decommissioned, and from unnecessarily burdening electric ratepayers with additional costs beyond that already paid the DOE in the Nuclear Waste Fund. However, to ensure timely and cost-effective removal, the Congress must clearly direct the DOE to accelerate removal from shutdown reactor sites.

Currently, the total amount of spent fuel and waste at permanently closed civilian reactors is 1,147 MTU, a small amount compared with the total amount of commercial spent fuel presently stored in the U.S. (approximately 32,000 MTU).

S. 1271 acknowledges the importance of removing the small volume of spent fuel and waste from shutdown reactor sites on a priority basis in Section 205(d)(3), which requires the DOE to accept spent fuel and high-level waste, including Greater-Than-Class-C waste (GTCC waste) from permanently shutdown reactors once the DOE has achieved the annual acceptance rate executed in the Standard Contract. Yankee fully supports this Section and the inclusion of GTCC waste. Yankee proposes that consideration be given to amending this section to require the DOE to accept an amount no less than 25% of the difference between the annual contract acceptance rate and the acceptance rate established in Section 507(a) of the bill. This change would ensure that the spent fuel currently stored at shutdown reactor sites, which comprises a small fraction of the total spent fuel needing to be stored at the interim storage facility, will be removed on a timely and cost-effective basis.

In addition, Yankee fully supports Section 205(b)(2) which requires the DOE to consent to contract amendments that provide for reimbursement to contract holders for transportable storage systems, purchased by contract holders, if DOE determines that it is cost-effective to use the systems as part of the integrated management system.

CONCLUSION

There is unprecedented support in the Commonwealth of Massachusetts and the New England region for the comprehensive reforms provided in S. 1271. The following organizations representing the state and/or region's governors, legislatures, regulators, businesses and labor organizations have all gone on record this year in support of legislation amending the NWPA: The New England Governors' Conference Power Planing Committee, the New England Conference of Public Utility Commissioners, the New England Council, the Massachusetts Legislature, the Greater Boston Chamber of Commerce, and the Massachusetts AFL-CIO. All these groups echo the need for the Senate to take action this session to resolve the nuclear waste storage situation directly impacting the state's and region's electric consumers and economy.

Yankee Atomic Electric Company supports S. 1271 and urges that the needs of permanently shutdown reactors be fully incorporated within the scope of this comprehensive legislation.

○